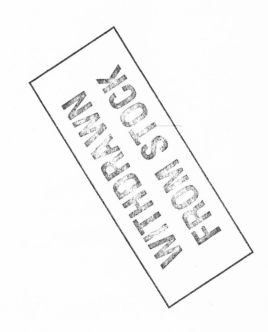

THE MOLECULAR BIOLOGY
OF DEVELOPMENT

THE MOLECULAR BIOLOGY OF DEVELOPMENT

BY

JAMES BONNER

California Institute of Technology

CLARENDON PRESS OXFORD

1965

Oxford University Press, Amen House, London E.C.4

GLASGOW   NEW YORK   TORONTO   MELBOURNE   WELLINGTON
BOMBAY   CALCUTTA   MADRAS   KARACHI   LAHORE   DACCA
CAPE TOWN   SALISBURY   NAIROBI   IBADAN   ACCRA
KUALA LUMPUR   HONG KONG

© Oxford University 1965

Printed in Northern Ireland at The Universities Press, Belfast

PREFACE

THE time has come for direct attack upon the central problem of biology, the problem of how it is that a single cell, the fertilized egg, gives rise to an adult creature made of many different kinds of cells. This process, which we know as development, has been described and thought about by biologists for as long as there has been a science of biology. Its nature has remained a mystery because we have not heretofore understood enough about the nature of life itself. Today we do. We know that all cells contain the directions for all cell life, written in the DNA of their chromosomes, and that these directions include specification of how to make the many kinds of enzyme molecules by means of which the cell converts available substances into metabolites suitable for the making of more cells. This is the picture of life given to us by molecular biology and it is general, it applies to all cells of all creatures. It is a description of the manner in which all cells are similar. But higher creatures, such as people and pea plants, possess different kinds of cells. The time has come for us to find out what molecular biology can tell us about why different cells in the same body are different from one another, and how such differences arise.

To this new study of development biologists and molecular biologists alike can and must contribute. This book is addressed to the biologist who wishes to acquaint himself with those portions of molecular biology which are pertinent to the study of development. It is addressed, too, to the molecular biologist who wishes to acquaint himself with some of the ways in which the insights of modern biology are being applied to developmental matters. This is a small book and it cannot therefore include all of the facts pertinent to both developmental and molecular biology. It is the author's hope, however, that it can serve as a guide to a new land, a guide to the new world of the molecular biology of development.

This book has been written at Oxford during my tenure of the Eastman Visiting Professorship. My gratitude is due to all those who made my stay in Oxford such a pleasant and productive one; Professor Geoffrey Blackman for the hospitality of his department; Professor Hans Krebs for the hospitality of his laboratory; Dr. Kenneth Burton for many fruitful discussions; and my colleagues of Balliol College for many pleasant hours spent in their company. My thanks are due, too, to Dr. Michael Bary and Dr. Daphne J. Osborne who have read and criticized the manuscript. Finally I wish to acknowledge with affection and thanks my colleagues of the California Institute of Technology whose discoveries and whose stimulation over the years have generated and maintained my own interests in developmental matters.

CONTENTS

1

THE MOLECULAR FRAMEWORK
OF DEVELOPMENT

BIOLOGY is today in the midst of a knowledge explosion. During the past quarter century and more particularly during the past ten years we have gained more and deeper insight into the logic and nature of life than in all the previous history of our science. As a result, it is now both more important and more difficult than ever before to ask and to answer the following questions. What are the important remaining problems of biology? Which of these problems are fruitfully approachable? How can we use the framework of what we now know as molecular biology to help us to approach and solve the previously unapproachable and insoluble? These are the questions of biology today to which I wish to address myself.

Let us first consider the framework of modern biology. I believe it can be truly said that the knowledge explosion in biology commenced when biologists began to concentrate their attention on cells rather than on whole creatures and, moreover, ceased looking at the differences between different kinds of cells and concentrated instead on their similarities. This point of view first took hold among the enzymologists. An enzymological acquaintance of mine, when recently upbraided by a biologist for his rudimentary knowledge of the different types of cells and indeed of different kinds of creatures, replied cheerily, 'Well, I do not need to be able to tell them apart; they all look the same in a Waring blendor'. The enzymologist takes some cells and grinds them up and finds an enzymatic activity and the first thing we know we have modern enzymology. This approach started with Harden and Young and continued with Emden, Michaelis, Meyerhof, and Lohman. From the early work on glycolysis and the quantum jump in insight provided by Krebs and by Lipmann have developed modern concepts of metabolic pathways, the pathways along which carbon and other atoms are switched around to form the essential building blocks of life. Today we know a great deal about these pathways in living things. We know generally which kinds of enzymatic transformations are permissible and which proscribed. We can predict which new kinds of metabolic transformations to search for. The study of metabolic pathways has become a sort of chess with a well recognized board upon which well recognized pieces may be moved only according to well recognized moves. True, there are great gaps in our knowledge. We do not know exactly how oxydative phosphorylation occurs nor do we know how membranes spin themselves or are spun, but we can foresee the day when these things will be known. So much for enzymology and the pathways of metabolism. They have become neoclassical biology.

A second great development which has also become neoclassical biology, concerns the path of information flow; information storage, information transfer, information readout, all on the molecular level. It is the lore of information flow amongst the molecules to which we now give the name of molecular biology. Let us review the state of our knowledge of molecular biology. If we take a cell, any cell, and look at it, we find that it contains a variety of subcellular particles. It contains a nucleus (generally one); some chloroplasts, fifty or so (but only in plants); mitochondria, five hundred or so; ribosomes, five hundred thousand or so; and enzyme molecules, five hundred thousand thousand or so. One of the great triumphs of biology has been the separation of these subcellular entities from one another and the consequent discovery of what each is made of and what each contributes to the overall cellular welfare. The enzyme molecules of a few thousand different kinds catalyze the several thousand different kinds of chemical reaction which convert nutrient molecules into the molecules from which the subcellular structures are built. Mitochondria and chloroplasts are both basically concerned with enzyme matters. Each constitutes a pre-packaged assembly of those kinds of enzyme molecules needed for a particular set of metabolic transformations.

Enzyme molecules are protein molecules. Each species of enzyme molecule consists of a long chain (or a few such chains) composed of the twenty different kinds of amino acid molecules stapled together in a specific and unique sequence. If we wish to make a particular kind of enzyme molecule, we must have information; information about which kind of amino acid molecule to insert next as we assemble the growing peptide chain. Enzyme synthesis is therefore an information-requiring task and in the cell this task is entrusted to the ribosomes, of which the essential information-containing component is the long punched tape which contains, in coded form, the instructions concerning which amino acid molecule to put next to which in order to produce a particular enzyme. We shall see later that the ribosomal tape contains a series of messages. The first says in effect something like 'To read the message which I contain, start at this end' and the last says in effect 'End of message, mission accomplished, release the enzyme molecule and send it out to work'. In between lie the mystic rituals of protein synthesis which are, however, sufficiently well understood to have become a part of neoclassical enzymology. The essential and interesting aspect of protein synthesis is the punched tape, for which the ribosomal particle is in a sense the decoder. This tape is, of course, a molecule of RNA, a long molecule made of the four kinds of RNA monomers. We know that information is contained in this messenger RNA molecule in the sequence in which the four monomers succeed one another down the chain. Messenger RNA is written in a four-letter alphabet and in it are encoded words about which of the twenty amino acid molecules to place next to which to produce some particular species of enzyme molecule. The way in which the twenty different letters of the amino acid alphabet are represented in the four letter alphabet of RNA is known as the coding problem. My colleague, George Wells Beadle, in his Eastman lectures (1958–1959) paid a great deal of attention to the coding problem, devising various possible codes for the desired translation. Since that time there have been great strides in the understanding of coding matters. There occurred in 1961 what we now refer to as the U3 incident which was followed by the great coding war of 1962. As a result of the spirited activities of Nirenberg and Matthaei and of Ochoa and his colleagues, we now know a great deal about how the several amino acids are represented in RNA language and we may be assured that all will be known in the near future (summarized in Crick, 1963). Coding too is passing into neoclassical biology.

So the enzyme molecules are made by ribosomes using messenger RNA tapes as their source of information. The only remaining question, then, is: where does the messenger RNA come from; how does the messenger RNA get information put into it? The answer to this question is simple: the messenger RNA is made by the genetic material of the cell, by the DNA, using itself as template. We have in a sense known this since 1941 when George Wells Beadle and Edward Leroy Tatum showed that for each enzyme in the cell there is a gene in the genetic material which specifies that enzyme. Or, to put it the other way round, that the function of each gene is to supervise the production of a particular kind of enzyme. 'One gene:one enzyme' was the rallying cry of geneticists in those far off days. But since geneticists did not then know about messenger RNA and ribosomes, they did not know that it is these entities which are directly responsible for enzyme synthesis. Today we know that one gene—one messenger RNA—one enzyme is the true sequence of information transfer in the living cell. This is really all there is to the framework of molecular biology. The DNA sits in the cell in double-stranded, base-paired structure, A holding hands with T, G holding hands with C, and it prints off single-stranded copies of itself, copies made of the RNA monomeric units. These are the messenger RNA molecules which then go forth to cause, with the co-operation of the ribosomes, the assembly of enzyme molecules. The enzyme molecules transform whatever materials are available into building blocks for making more enzyme molecules and more messenger RNA molecules. But all of this machinery is contained in the cell so that there can be a few further kinds of enzyme molecules which catalyze the reactions by which DNA building blocks are made. For the DNA has one further capability, the most powerful attribute of all. DNA can replicate itself, provided only that the appropriate building blocks are present. After the DNA has replicated itself the cell can divide into two cells. The logic of the living creature as it is seen by molecular biology consists in just this tricycle of life.

This then is the framework of molecular biology, the framework within which we believe all cells are similar. Much remains to be done by way of upholstering the framework but basically it is now a part of neoclassical biology. What is to be done next? Of what can modern biology consist? It is my thesis that we can now profitably go back and re-examine what it is that makes the differences between the different kinds of cells of a higher organism. We can take the new look within the framework of molecular biology, using this framework as the guide and key to new hypothesis, new kinds of experiments, new insights. We can look again at differentiation and at how creatures respond to external stimuli. We can perhaps in one limited sense look with new eyes at the biology of memory and of learning. Of these problems, it is that of differentiation to which we will address ourselves.

There are many ways, of course, in which to approach differentiation. We can, for example, merely note that such a phenomenon exists. The descriptive study of embryology is the classical path. What can we say, however, about differentiation in the light of our knowledge of molecular biology? First, we can say that every cell of the higher organism meiosis, polyploidy, and polyteney apart, contains the same amount and kind of DNA. Whether or not each cell contains the whole genome of the organism is in fact a fact has been debated in some quarters but not amongst the plant biologists who know it to be true. We have in plant biology a substantial number of examples of a single specialized cell of the adult organism being brought to renewed cell division and thence to develop through a whole life cycle into an adult organism. The entire genetic complement is thus clearly contained in the differentiated cell. But though each specialized cell contains the whole genomal DNA, the different kinds of specialized cells do differ in the amounts and kinds of proteins which they contain. A classical example is the erythrocyte and its precursor, the reticulocyte, which produce principally a single protein, haemoglobin. There are in the higher animal genes for making haemoglobin. We may conclude that in the erythrocytes and their precursor cells the genes for making haemoglobin are turned on, to produce the messenger RNA for the production of haemoglobin. Contrarywise, in other specialized cells of the same organism no haemoglobin is produced although other enzymes are. Our new knowledge of molecular biology tells us that in these other cells the genes for making haemoglobin are inert, do not make their messenger RNA. This sort of example might be multiplied many times. Each kind of specialized cell of the higher organism contains its characteristic enzymes but each produces only a portion of all of the enzymes for which its genomal DNA contains information. Clearly then, the nucleus contains some further mechanism which determines in which cells and at which times during development each gene is to be active and produce its characteristic messenger RNA, and in which cells each gene is to be inactive, to be repressed.

The egg is activated by fertilization (itself a subject suitable for study within the framework of molecular biology) and divides into two cells and then into four cells. As division proceeds cells begin to differ from one another and to acquire the characteristics of specialized cells of the adult creature. There is then within the nucleus some kind of programme which determines the property sequenced repression and derepression of genes and which brings about orderly development. What is the mechanism of gene repression and derepression which makes possible development? Of what does the programme consist and where does it live? We can say that the programme which sequences gene activity must itself be a part of the genetic information since the course of development and the final form are heritable. Further than this we cannot go by classical approaches to differentiation. There are, however, fresh paths along which we may attack the problem.

Let us then go to specifics. We will select a particular gene of a particular higher organism and the product protein of that gene; a protein which is characteristically made in particular cells during a particular part of the developmental cycle of the organism in question and not produced in other cells during other parts of the developmental cycle. We may remark in passing that this kind of control of genetic activity, the appearance in particular cells of a particular protein during particular parts of the life cycle, can be studied only in higher organisms, since such control appears to be absent among bacteria. The latter, on the contrary, possess only inductive or feedback control of genetic activity. My colleague, Ru-chih C. Huang and I have chosen as an example of a protein subject to developmental control, the reserve globulin of pea seeds. This protein is made and stored in developing pea cotyledons and is not made in other tissues, as is shown in the data of Table 1. The manufacture of pea-seed globulin is controlled by specific genes. So, to study why some cells of the pea plant make reserve globulin and others do not, we must focus our attention on the genetic material of the cells involved. Our first sub-task is to isolate the genetic material. Let us therefore consider the lore of chromosome husbandry.

TABLE 1

Synthesis of pea-seed globulin by varied organs of the pea plant

Organ†	C^{14}-leucine incorporated into protein		Globulin/total protein (%)
	Total soluble protein (cpm)	Globulin (cpm)	
Flower	17500	510	2·9
Cotyledon	7100	337	4·7
Older Cotyledon	9200	862	9·3
Root	17150	31	0·18
Apical bud	11600	14	0·12
Apical bud	27600	42	0·15

† The different organs were separately incubated in L-leucine (1 μc/ml, 19 μc/μm) for 2 hours at 25° in the presence of penicillin (6 μg/ml). The soluble protein was obtained by dialysis of the supernatant of tissue homogenate which had been centrifuged at 105000 × g to remove all particulate material. Pea-seed globulin was determined by an immunochemical detection system (after Bonner, Huang, and Gilden, 1963). The background of the detection system is about 0·13 per cent, that is, the system detects about 0·13 per cent globulin in protein mixtures which contain no globulin.

The isolation of chromosomes from the cells of higher creatures has proven to be surprisingly simple. All one has to do is to grind tissue in such a way as to tear open the cell and nucleus membranes. One then filters off the membranes and centrifuges the grindate in a centrifugal field too slight to spin down mitochondria. The chromosomes are the biggest and heaviest subcellular structures in the homogenate (apart from starch grains) and so they sediment first. In this way it is possible to recover 95 per cent or more of the tissue DNA as crude chromosomal material, contaminated of course with nucleolar fragments, some aggregated mitochondria, and so on.

We then purify this crude chromatin by layering it over 1·9 molar sucrose and centrifuging it through the density gradient thus established. The chromatin sediments through the denser sucrose solution while the protein contaminants remain in the supernatant. In this way purified chromatin of chracteristic chromosomal properties and composition can be readily obtained from a variety of cells and tissues.

Chromosomal material, chromatin isolated by the methods outlined above, possesses an interesting and important biological activity; it is able to synthesize RNA from the four riboside triphosphates. Such synthesis depends in the first place upon the simultaneous presence of the four triphosphates. If any one of them is left out, RNA synthesis does not occur, nor does it occur if the riboside diphosphates are used. The mechanism of chromosomal RNA synthesis is then different from the RNA phosphorylase reaction described by Ochoa and his colleagues some years ago, which specifically requires the diphosphates as substrates. The synthesis of RNA by chromatin is of the type which we now know as DNA-dependent RNA synthesis; the assemblage of RNA by chromatin is dependent upon the presence in the chromatin of intact DNA molecules. Brief exposure of chromatin to DNA-ase results in total destruction of its ability to support RNA synthesis. Finally the ability of chromatin to synthesize RNA depends on an enzyme since it is destroyed by brief heating of chromatin at 60°. This enzyme, we name RNA polymerase. We shall not digress to present the rigorous evidence that the material produced by chromatin is in fact RNA. We shall merely note that the amount of RNA generated by chromatin is vastly increased in the presence of supplementary RNA polymerase (Chapter 3). Let us go directly to the biological properties of the RNA thus chromosomally generated. Its most important property is that it is capable of supporting the synthesis of protein by ribosomes which have been appropriately freed of their own messenger RNA.

We are now then in a position to determine whether chromatin isolated from the living cell preserves the control of genetic activity characteristic of life. That it does so is shown in Table 2. For the experiments of Table 2, we first isolate chromatin from pea buds, a tissue of the pea plant which does not make pea-seed globulin in life. Such chromatin supports RNA synthesis and the RNA so synthesized supports protein synthesis by our ribosomal system. But the protein thus synthesized does not include any significant proportion of pea-seed globulin. This then is quite as it is in the living cell. Next, we isolate chromatin from developing pea cotyledons, the organ of the pea plant which in life makes pea-seed globulin in abundance. The chromatin of pea cotyledons supports RNA synthesis and the

TABLE 2

Synthesis of pea-seed globulin by messenger RNA dependent ribosomal system in response to messenger RNA generated by two different kinds of pea-plant chromatin

Template for RNA synthesis†	C^{14}-leucine incorporated into protein		Globulin/total protein (%)
	Total soluble protein (cpm)	Globulin (cpm)	
Apical bud chromatin	15650	16	0·10
Apical bud chromatin	41200	54	0·13
Cotyledon chromatin	8650	623	7·2
Cotyledon chromatin	6500	462	6·9

† The reaction mixture contains all materials required for both RNA and protein synthesis. Incubation for 30 minutes at 37°. All particulate material was then centrifuged off at 105000 × g and pea-seed globulin content of soluble protein synthesized determined by immunochemical assay.

RNA so synthesized supports protein synthesis by the ribosomal system, and the protein so synthesized includes the pea-seed globulin as a major component. This also then is as it is in life.

The control of genetic activity which characterizes cells of the pea plant in the living plant is then preserved in isolated chromatin. We have therefore a system to which it is possible to apply the tools of biochemistry and biophysics to find out what it is that controls genetic activity. We may try to find how to turn on the globulin-making genes in pea-bud chromatin in which these genes are repressed, or to turn off the globulin-making genes in pea-cotyledon chromatin in which these genes are active. We can study the control of genetic activity, not by genetic machinations, but by molecular biology. It is with the control of genetic activity as studied on the molecular level that the next few chapters will be concerned.

SELECTED REFERENCES

Coding

CRICK, F. H. C., *Progress in Biophysics and Biophysical Chemistry* (1963).

One gene—one enzyme

BEADLE, G. W. and TATUM, E. L., *Proc. Nat. Acad. Sci., Wash.* **27,** 499 (1941).
BEADLE, G. W. in *The Chemical Basis of Heredity*, p. 3. Johns Hopkins Press (1957).

Totipotency of the plant cell and related matters

SWIFT, H., *Proc. Nat. Acad. Sci., Wash.* **36,** 693 (1950).
STEWARD, F. C., MAPES, M. O. and KENT, A. E., *Am. J. Bot.* **50,** 618 (1963).
EARLE, E. D. and TORREY, J. G., *Am. J. Bot.* **50,** 614 (1963).
BRAUN, A. C., *Proc. Nat. Acad. Sci., Wash.* **45,** 932 (1959).
ZEPF, E., *Bot.* **40,** 87 (1952).

Chromosome isolation and chromosomally supported RNA *and protein synthesis*

HUANG, R. C. C., MAHESHWARI, N. and BONNER, J., *Biochem. biophys. Res. Commun.* **3,** 689 (1960).
BONNER, J., HUANG, R. C. C. and MAHESHWARI, N., *Proc. Nat. Acad. Sci., Wash.* **47,** 1548 (1961).
BONNER, J. and HUANG, R. C. C., *Can. J. Bot.* **40,** 1487 (1962).
HUANG, R. C. C. and BONNER, J., *Proc. Nat. Acad. Sci., Wash.* **48,** 1216 (1962).
BONNER, J., HUANG, R. C. C. and GILDEN, R., *Proc. Nat. Acad. Sci., Wash.* **50,** 893 (1963).

Other general summaries of the study of development

MCELROY, W. and GLASS, B., *The Chemical Basis of Development.* Johns Hopkins Press (1958).
WADDINGTON, C. H., *New Patterns in Genetics and Development.* Columbia University Press (1962).
RUDNICK, D. (Ed.), *Developing Cell Systems and their Control.* Ronald Press (1960).

2

CHROMOSOME HUSBANDRY

WE HAVE seen in Chapter 1 that chromatin as it is isolated from the cell can synthesize messenger RNA, and synthesizes in fact the kind of messenger RNA which is made by that same chromatin in life. The particular genes upon which we have focused our attention are those for making messenger RNA which directs the synthesis of pea-seed globulin. They are turned on, or derepressed, in chromatin from cells which in life synthesize pea-seed globulin and are turned off, or repressed, in chromatin from cells which in life do not synthesize pea-seed globulin. Our next problem is to discover what elements of the chromatin exert such control of genetic activity.

Let us further consider the isolation and the properties of chromatin. For these studies we have used chromatin isolated from the embryonic axis of pea seedlings. Pea seeds are caused to germinate and the embryonic axes removed from the cotyledons when they are about 1 cm long. At this stage the cells of the embryo are small and richly filled with nuclei, and best of all, this embryonic tissue is available in large quantities. Chromatin isolated from pea embryos by the methods outlined in Chapter 1 has the composition indicated in Table 1. It

TABLE 1

Composition of purified pea-embryo chromatin and of the nucleo-histone prepared from it†

Property	Purified chromatin	Nucleohistone from same
DNA: % of mass	36·5	41·5
Histone: % of mass	37·5	55·0
Histone/DNA ratio	1·03	1·32
RNA: % of total nucleic acid	28	5·4
Nonhistone protein: % of total protein	20	0

† After Bonner, J. and Huang, R. C. C., 1963.

consists of DNA, some RNA and some associated protein. The protein is of two principal types. The great bulk of it belongs to that group of proteins known for over a hundred years as histones; basic proteins containing a high proportion of the positively charged groups of lysine and arginine. In pea-embryo chromatin, histone and DNA are present (on a mass basis) in approximately equal amounts. Chromatin further contains non-histone protein, and this includes, as we shall see below, the RNA polymerase and other enzymes of the genetic material.

We then consider the separation of chromatin into its constituents; into the DNA and protein which make up its bulk. For this purpose we use a method as simple as it is elegant. The chromatin is dispersed in 4 M cesium chloride (CsCl). At the ionic strength of 4 M cesium chloride all ionic bonds within the chromatin are disrupted, and since the protein is largely bound to the DNA by ionic bonds, protein and DNA separate. At the density of 4 M cesium chloride, about 1·4, nucleic acid sinks while protein floats. We therefore homogenize chromatin in 4 M cesium chloride and centrifuge it briefly. The protein forms a skin on the surface, the nucleic acid pellets. The proteins of the skin include both histones and non-histones. Non-histone protein is separated from histone by extraction of the skin with dilute (0·05 M) tris buffer, pH 8·0, in which only non-histone proteins are soluble.

The first interesting fact which emerges is that exposure to high ionic strength abolishes all chromosomal structure. The DNA recovered from the pellet possesses a sedimentation coefficient of eighteen Svedberg units, corresponding to a molecular weight of approximately nine million. This molecular weight is similar to that found for the DNA of other organisms similarly prepared by deproteinization. As in other cases too, the pure DNA prepared from pea chromatin is almost certainly degraded, broken into pieces smaller than those present in the chromosome.

We are now in a position to purify the chromosomal RNA polymerase, the enzyme which polymerizes ribonucleotides to RNA. This enzyme is found in the non-histone portion of the chromosomal protein and is in part released from the chromosomal DNA by 4 M cesium chloride. The enzyme has been purified by extraction of the chromosomal proteins with neutral tris buffer followed by protamine precipitation and fractional ammonium sulphate extraction of the precipitate. The thus purified enzyme supports RNA synthesis provided only that DNA and the four riboside triphosphates are present. The data of Fig. 1 show that the DNA requirement is saturated by a DNA concentration of approximately 50 μg per 0·5 ml of

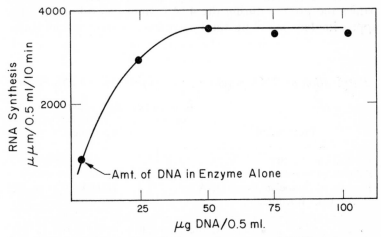

FIG. 2.1. Dependence of purified pea chromosomal RNA polymerase upon added pea DNA. Tested in complete reaction mixture for RNA synthesis.

reaction mixture. The enzyme itself, even purified preparations, contains endogenous DNA in small amount, and electron-micrographs reveal that the enzyme preparation consists of small chains of DNA with a blob, presumably protein, at one end. For this reason the chromosomal RNA polymerase has a low dependency ratio; a low ratio of RNA synthesizing activity in the presence and absence of added DNA. It is of the order of 5 or 7 to 1. Nonetheless, the data of Fig. 1 provide strong evidence that the synthesis of RNA by chromatin is DNA-dependent. That DNA-dependent RNA synthesis in fact involves transcription of the DNA molecule has been shown first by Furth, Hurwitz, and Goldmann (1961) who used as template synthetic poly dAT (polydeoxyadenylate-thymidilate copolymer) in which A and T alternate with one another along the DNA chain and showed that the RNA produced by RNA polymerase is poly AU (poly adenylate-uridilate copolymer) in which A and U alternate with one another along the RNA chain. We hold therefore as an article of dogma, although not yet fully proved, that the transcription of the DNA message by RNA polymerase involves the copying of the DNA template by complementary base-pairing methods.

Although part of the RNA polymerase of chromatin accompanies the protein during the 4 M cesium chloride treatment, approximately half of the total polymerase activity accompanies

TABLE 2

Purification of RNA polymerase from pea-embryo chromatin

Preparation	RNA synthesizing activity $\mu\mu$m nucleotide incorp. into RNA/10 min per mg protein
Purified chromatin	1100
Solubilized enzyme†	15800
Enzyme still associated with DNA after CsCl fractionation	118000

† Tested in presence of added DNA. After Huang, R. C. C. and Bonner, J., 1962.

the DNA to the pellet. This portion of the polymerase is apparently not bound ionically to DNA but bound in some firmer fashion. This separation of the DNA bound enzyme from other proteins by 4 M cesium chloride provides an important purification of polymerase activity as is shown in Table 2. By this one step the enzyme is enriched over a hundredfold on a protein basis.

Enzymologists always boast about how highly they have purified the enzyme in which they are interested. Let us consider the purification of pea-embryo chromosomal polymerase. The polymerase is purified one hundredfold from the level of purified chromatin. Purified chromatin has, however, been purified from pea embryos. One kilogram of pea embryos yields approximately fifty milligrams of chromosomal protein. This is in itself a purification of twenty thousandfold. One hundred times twenty thousand equals two million. Pea embryos however, have been purified from whole pea seedlings. 12·5 kg of pea seeds yield 1 kg of pea embryos, a purification of 12·5-fold. 12·5 times two million gives twenty-five millionfold for the overall purification of chromosomal RNA polymerase!

Obviously if we are to work with a component of tissue which is to be purified twenty-five millionfold, we must have a substantial amount of starting material. It is inadequate to germinate pea seeds in trays and pick off the embryos by hand, a procedure which yields about 10 g of tissue per person-hour. We have evolved a simplified production of pea embryos. In our new procedure, 25 lb of dry pea seeds are placed in a 35-gallon barrel. The barrel has a drain hole in the bottom and is mounted on castors for easy movement. The drain hole is first stoppered, the barrel filled with water, and left for 5 hours. This allows the seeds to swell. The stopper is then removed and the barrel pushed under a spray nozzle which sprays water at 20° over the peas, a temperature good for their germination. The spray water drains gently down through the seeds and out through the hole at the bottom, washing away any bacteria etc. which may begin to grow. After 24 hours at 20° the pea seeds have germinated. The problem is now to separate the embryonic axes from the cotyledons. To this end we have developed what we call a semi-automatic three-stage disassembly line. The germinating seedlings are first scooped from the barrel into the hopper of a pea mill. The mill has two rapidly counter-rotating rollers, fluted so as to encourage pea seeds to pass between them. They are set about 4 mm apart so that the embryonic axes will not be crushed. As the pea seedlings pass at high speed through the rollers they experience some shear, on the average enough to break both cotyledons from the growing embryonic axis. The effluent from the pea mill is then a mixture of pea cotyledons and pea embryonic axes. For their separation from one another a household electric washing-machine is used. The pea mill is mounted directly over the washing-machine and the pea seeds are ground directly into it. The water in the machine contains sucrose sufficient to raise its density to 1·035, a density at which pea embryonic axes float while pea cotyledons sink. The embryonic axes are simply scooped off the surface of the solution in the washing-machine. But the pea embryos thus collected are contaminated with pea seed-coats which are also partly liberated by the pea mill. For the final separation of pea seed-coats from embryos a slotted bed shaker is used. The slots in the bed of this machine are cleverly designed to let pea embryonic axes through yet retain pea seed-coats: the axes are washed through and collected in a wire basket. They are the starting material for our large scale chromatin preparation. This procedure provides abundant starting material with but little labour. For the study of chromatin of other cells and tissues, it may also be desirable to develop analogous semi-automated systems for providing large amounts of starting material.

TABLE 3

Ability of chromosomal DNA to support RNA synthesis as related to removal of chromosomal histone

Preparation	RNA synthesized/10 min $\mu\mu$m nucleotide/mg DNA
Crude chromatin	1175
Purified chromatin†	1175
Deproteinized DNA plus chromosomal polymerase	90000

† Prepared from crude chromatin by sucrose density gradient centrifugation.

We have discussed one of the principal components of the chromosomal RNA-forming system, RNA polymerase. Let us now consider the other, the chromosomal DNA. The DNA of chromatin is rather ineffective in the support of RNA synthesis relative to an equal amount of deproteinized DNA. Table 3 shows that while DNA of crude and purified chromatin is equally effective in supporting RNA synthesis an equal amount of pure DNA (fortified with purified soluble RNA polymerase) is far more effective. Perhaps however the comparison is not a fair one: perhaps the polymerase/DNA ratio in the artificially reconstituted system is higher than that in chromatin. To study this further, we have carried out the experiment outlined in Table 4. An aliquot of chromatin was fractionated with 4 M cesium chloride. From the protein skin thus obtained histone proteins were removed and the remaining proteins, including RNA polymerase) added back to the nucleic acid pellet. The ability of this reconstituted chromatin, now free of histone, to support

RNA synthesis was then compared with that of the original chromatin. The data of Table 4 show that rate of RNA synthesis per unit weight of chromosomal RNA is increased approximately fivefold as a result of histone removal. We may conclude therefore, that chromatin as it is obtained from the pea embryo, contains an inhibitor which causes the chromosomal DNA to be relatively ineffective in the conduct of DNA-dependent RNA synthesis. The experiment of Table 4 points the finger at the histone component as the responsible agent.

TABLE 4

The RNA synthesizing activity of pea-embryo chromatin as related to removal of histone

System	RNA synthesis/10 min $\mu\mu$m nucleotide incorp/mg DNA
Purified chromatin	450
Solubilized chromatin: histone removed†	2090

† Dissociated in 4 M CsCl. Nonhistone protein recombined with DNA fraction.
After Huang and Bonner, 1962.

Histones occur only in chromatin, and, thus far, only in those creatures which contain proper chromosomes. Generally speaking, indeed, histones are found only in organisms in which differentiation into different types of specialized cells is found. In such creatures, histone is bound to DNA as nucleohistone. In the nucleohistone component of chromatin the cationic groups of histone are equivalent to, and neutralize the anionic phosphate groups of the DNA. Histone is thus distributed over the entire length of the DNA chain, making it thicker than the double helix of DNA, approximately 35 Å in width as compared to 20 Å for DNA itself.

From the compositions of histone and of DNA we can calculate that in a nucleohistone in which DNA is fully complexed with basic protein, the mass ratio of histone to DNA should be approximately 1·35 to 1. In our pea-embryo chromatin, as we have seen, the mass ratio of histone to DNA is approximately

1 to 1. It appears therefore, that not all of the DNA in pea-embryo chromatin is fully complexed with histone. Our next sub-task will be the separation of that DNA which is fully complexed with histone, the nucleohistone component, from that DNA which is not fully complexed. Fortunately, this matter has been previously studied, first by Zubay and Doty in 1959. Their methods have been used by Peacock in his physical studies of nucleohistone, and we have used them in modified form for the separation of nucleohistone from chromatin. These methods are summarized in Fig. 2. Chromatin,

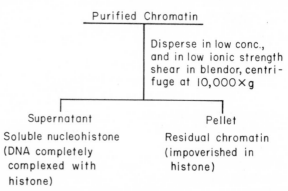

Fig. 2.2. Procedure for preparation of soluble native nucleohistone from purified chromatin.

dispersed at low concentration (0·5 to 1·0 mg DNA per ml) and at low ionic strength (0·004 M NaCl at neutral pH) is briefly sheared at low temperature in a homogenizer. By the shearing, the long stiff rods of nucleohistone are broken from the chromosomal mass, while the DNA molecules not completely covered with histone are not sheared (or less sheared) and remain interconnected in the chromosomal mass. The sheared chromatin is next stirred at low temperature to allow the sheared and unsheared elements to untangle. The whole is then centrifuged in a centrifugal field sufficient to bring down the chromosomal residues, leaving the soluble nucleohistone in the supernatant. Under these circumstances some 70 to 80 per cent of the initial DNA of pea-embryo chromatin remains in the supernatant. This DNA is fully complexed with histone as is shown in Table 1. The pelleted material on the contrary is impoverished in histone. The data of Table 1 show also that non-histone protein is absent from the nucleohistone component and that a portion of the RNA of the chromatin accompanies the nucleononhistone fraction rather than remaining in the pellet.

Solubilized nucleohistone prepared in this way has been studied extensively with respect to state of dispersal and other physical properties. It has been shown by Zubay and Doty (1959), by Giannoni and Peacock (1963), and by ourselves, that this soluble nucleohistone consists of DNA molecules of sedimentation coefficient 18 to 22 S (18 to 22 Svedberg units) complexed with histone to yield nucleohistone of sedimentation coefficient 27 to 32 S. The soluble nucleohistone consists then of molecularly dispersed individual DNA molecules each fully complexed with histone. A further feature of nucleohistone is the fact that the presence of histone in it stabilizes the DNA of the complex against melting. In DNA the bases are stacked one above another like poker chips. Due to the π-π interaction between them the absorption of ultraviolet light by the assemblage is less than that of an equivalent amount of unstacked, free, nucleotides. DNA is therefore said to be hypochromic. When a solution of DNA is gradually warmed, a melting temperature is ultimately attained at which the hydrogen bonding and hydrophobic stacking interaction between bases is disrupted. At this temperature the structure of the double stranded DNA molecule collapses. The absorption of light by the system increases to that characteristic of an equivalent amount of free bases. We say that DNA, upon melting, exhibits hyperchromicity. Figure 3 compares the melting profile of pea chromosomal DNA with that of the same amount of DNA in the form of pea chromosomal nucleohistone. Pea chromosomal DNA like all DNA, melts, not sharply but over a range, and we characterize the transition range by the temperature, T_M, at which half of all hyperchromicity has manifested itself. The T_M of pea-embryo DNA in the particular salt concentration used (0·016 M) is 70°. When DNA is complexed with histone it exhibits a melting profile which is both shifted to higher temperatures and is less sharp than that of DNA itself. The T_M of pea-embryo nucleohistone is 84°. Considerable stabilization is thus conferred upon DNA by histone.

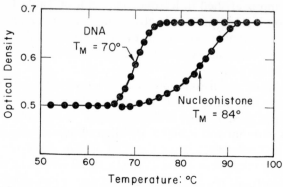

Fig. 2.3. Melting profiles of pea DNA and of pea DNA complexed as the native nucleohistone. Heating carried out in dilute saline citrate, 0·016 M.

Let us now ask, can DNA in the native nucleohistone complex serve as template for the support of DNA-dependent RNA synthesis? Native nucleohistone prepared as outlined is itself inactive in the conduct of RNA synthesis. Is this inactivity due to absence of RNA polymerase or to presence of histone? To test this matter, solubilized chromosomal RNA polymerase was added to nucleohistone. The results of such an experiment are outlined in Table 5. The data show that the chromosomal RNA polymerase used can synthesize RNA in the presence

TABLE 5

Relative inactivity of nucleohistone component of pea-embryo chromatin in support of DNA-dependent RNA synthesis by solubilized chromosomal RNA polymerase

Added to reaction mixture in addition to enzyme	RNA synthesis/10 min $\mu\mu$m nucleotide incorp. per mg enzyme
50 μg DNA	2030†
50 μg DNA as nucleohistone	190
50 μg DNA + 50 μg DNA as nucleohistone	1910

† Incorporation due to enzyme alone subtracted throughout.
After Bonner, J. and Huang, R. C. C., 1963.

of deproteinized pea DNA. They also show that native pea nucleohistone is essentially inactive in this function. The presence of nucleohistone does not interfere with the ability of the polymerase to utilize deproteinized pea DNA. The inactivity of native nucleohistone is not therefore due to interference with the activity of RNA polymerase.

Native nucleohistone is inactive or essentially so in the support of RNA synthesis over a large range of concentrations.

TABLE 6

Effectiveness of pea-embryo chromatin (heated to 60°
to inactivate RNA polymerase) in the support of
DNA *dependent RNA synthesis by added solubilized*
chromosomal RNA polymerase

Added to reaction mixture in addition to enzyme	RNA synthesized/10 min $\mu\mu$m nucleotide/mg enzyme†
50 μg DNA	2030†
250 μg DNA as whole chromatin (heated to 60°)	1600

† Incorporation due to enzyme alone subtracted throughout.
After Bonner, J. and Huang, R. C. C., 1963.

What can we say of chromatin? We already know that chromatin can support RNA synthesis under the auspices of its endogenous RNA polymerase. Can chromatin however use exogenous, added, RNA polymerase? Is our test of the activity of nucleohistone a fair one? To this end chromatin was briefly heated to 60° to inactivate its endogenous chromosomal RNA polymerase and the heated chromatin was then added to a reaction mixture containing solubilized RNA polymerase. In contrast to the behaviour of its nucleohistone component, whole chromatin does support RNA synthesis under the auspices of exogenous RNA polymerase. The data of Table 6 show that chromatin is roughly as active as one-fifth as much deproteinized DNA. It would appear that only a portion of the DNA of chromatin is present in active form.

The facts deduced above begin to provide us with a picture of one aspect of chromosomal structure. It would appear that a portion of the DNA of pea-embryo chromatin is present as nucleohistone, as DNA fully complexed with histone, and that this nucleohistone component is inactive, *in vitro* at least, in supporting RNA synthesis. A further portion of the DNA of chromatin appears to be present in different form, not complexed with histone, and active in supporting RNA synthesis. What further evidence can we adduce concerning this hypothesis? Strong evidence that the picture is correct is provided by the melting profile of chromatin. Such a profile (Fig. 4) shows

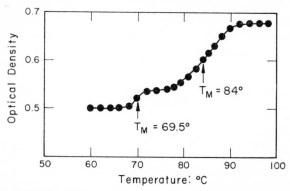

FIG. 2.4. Melting profile of purified pea-embryo chromatin. Heating carried out in dilute saline citrate, 0·016 M.

that the hyperchromicity of pea-embryo chromatin occurs in two steps. The T_M of the first step, approximately 70°, is close to that of pea DNA itself. The T_M of the second step, 84°, corresponds to that of nucleohistone. We may conclude, therefore, that in pea-embryo chromatin DNA is in fact present in two forms, one fully complexed with histone, the nucleohistone component, and one not thus complexed with histone. Of these two forms only the latter is active in the support of DNA-dependent RNA synthesis.

Next we shall determine whether it is possible to assign repressed genes to the nucleohistone component of the genome, derepressed genes to the histone not covered component. To do so we consider again the specific gene we have chosen, that which is responsible for the control of the formation of pea-seed reserve globulin. We recall that this gene causes a particular protein to be produced in developing pea cotyledons but not by other organs of the pea plant. We recall from Chapter 1 that RNA generated by chromatin of developing pea cotyledons causes the production by a ribosomal system of protein which includes a generous proportion of pea-seed globulin. RNA generated by chromatin from the vegetative buds of pea plants also supported the synthesis of protein but this protein did

TABLE 7

Synthesis of pea-seed globulin by messenger RNA-dependent ribosomal system in response to messenger RNA generated by pea-bud chromatin or by such chromatin after removal of histone

Template for RNA synthesis†	Total soluble protein (cpm)	globulin (cpm)	globulin/ total protein (%)
Apical-bud chromatin	15650	16	0·10
Apical-bud chromatin	41200	54	0·13
DNA of bud chromatin	15200	60	0·40
DNA of bud chromatin	14200	72	0·50
DNA of cotyledon chromatin	5600	22	0·39
DNA of cotyledon chromatin	60000	314	0·52

C^{14}-leucine incorporated in protein

† The reaction mixture contains all materials required for both RNA and protein synthesis. Incubation for 30 minutes at 37°. All particulate matter was then centrifuged off at 105000 × g and pea-seed globulin content of soluble protein synthesized determined by immunochemical assay. The background of the assay (no globulin present) is about 0·13 per cent.
After Bonner, J., Huang, R. C. C., and Gilden, 1963.

not include pea-seed globulin. We have concluded that the control of genetic activity characteristic of the cell in life is preserved, or largely so, in the isolated chromatin. Our task is then to discover what component of the chromatin exerts this control. Let us approach this question first by finding out how to derepress the gene for globulin-making in chromatin in which this gene is repressed. The data of Table 7 show that DNA prepared by deproteinization of pea-bud chromatin supports the synthesis of pea-seed globulin. True, the proportion of pea-seed globulin is not great, 0·2 to 0·3 per cent of total protein formed (above the background level of about 0·13 per cent), but is nonetheless quite detectable and real. The deproteinized DNA of pea-bud chromatin no longer contains any nucleohistone component. All its genes are therefore derepressed and support messenger RNA synthesis, the globulin-making gene among them. We expect of course to find the gene for globulin making in pea-bud chromatin, albeit in repressed form. After all, the bud will ultimately give rise to flowers and hence to ovules and pollen and hence to zygotes from whence will spring cotyledons which require the pea-seed globulin gene.

If the view that removal of protein including histone derepresses all genes of the genome is correct then such treatment of cotyledon chromatin should result in DNA which supports the production of additional species of messenger RNA which the native chromatin did not produce. The protein synthesized by such RNA, should then include many more kinds of protein than those whose synthesis is supported by pea-cotyledon chromatin itself. Hence the proportion of pea-seed globulin in the protein synthesized should be lowered by removal of histone from cotyledon chromatin. This expectation is fulfilled as shown by the data of Table 7. The proportion of globulin is in fact essentially the same, whether protein synthesis is supported by DNA of bud chromatin or by DNA of cotyledon chromatin. This again is to be expected, and is in accord with the thesis that all cells possess a common and complete set of genetic information.

It appears then, that we have discovered something. Histones are agents of gene repression. This is in itself moderately gratifying to know, giving as it does a cluelet to the hardware, the physical basis, of a genetic control mechanism. But our hard-won factlet poses at once a whole series of new questions; it is but the trunk of a problem tree. How many species of histones are there? Is there a different kind of histone for each gene which is to be repressed? Are the amino acid sequences of histones genetically determined? From whence do histones come? How are histones deposited upon and removed from the DNA? Are all repressors histones or other substances also involved? It is to these questions that later chapters will be addressed. In preparation for the consideration of these questions we must first, however, consider in more detail both the generation of messenger RNA and the process of protein synthesis.

SELECTED REFERENCES

Properties of chromatin and nucleohistone
BONNER, J. and HUANG, R. C. C., *J. mol. Biol.* **6,** 169 (1963).

Purification of chromosomal RNA *polymerase*
HUANG, R. C. C. and BONNER, J., *Proc. Nat. Acad. Sci., Wash.* **48,** 1216 (1962).

Transcription of poly dAT
FURTH, J. J., HURWITZ, J. and GOLDMANN, M., *Biochem. biophys. Res. Commun.* **4,** 431 (1961).

Preparation and properties of nucleohistone
ZUBAY, G. and DOTY, P., *J. mol. Biol.* **1,** 1, 1959.
GIANNONI, G. P. and PEACOCKE, A., *Biochem. biophys. Acta* **68,** 157 (1963).
BONNER, J. and HUANG, R. C. C., *J. mol. Biol.* **6,** 169 (1963).
BONNER, J. and Ts'o, P.O.P., Editors. *The Nucleohistones.* Holden-Day. (1964).

Histone and control of gene activity
BONNER, J., HUANG, R. C. C. and GILDEN, R., *Proc. Nat. Acad. Sci., Wash.* **50,** 893 (1963).

3

AMPLIFYING THE CHROMOSOMAL MESSAGE

WE HAVE seen that chromatin consists of DNA, some RNA, some RNA polymerase for the conduct of RNA synthesis, and of that characteristically chromosomal protein, histone. We have further seen that in chromatin a portion of the DNA is fully complexed with histone, a complex constituting the nucleohistone component of chromatin. A further portion of the genomal DNA is not so complexed, or not so completely complexed with histone. These two major subfractions of the genomal chromosomal component can be differentiated by their chemical and physical as well as by their biological properties. The biological property of relevance is ability to support DNA-dependant RNA synthesis. We have seen how it is possible to purify from chromatin the enzyme responsible for joining ribonucleotides together on the DNA template—the enzyme RNA polymerase. In the support of RNA synthesis by RNA polymerase, DNA fully complexed with histone is inert, or nearly so. Clearly the properties of RNA synthesis which are possesed by chromatin as a whole are to be attributed to a component which is not complexed with histone in the manner characteristic of nucleohistone.

We will now make a digression, but one which will lighten our subsequent tasks. We can sense that the key to chromosomal activity is the enzyme RNA polymerase. RNA polymerase is, in a very real sense, the readout device for the genetic information. It prints out in usable form as messenger RNA, information contained in genes, but only in those genes which are derepressed.

We have seen however, the difficulties which attend the use of chromosomal RNA polymerase; it is hard to get a great deal of it, and it is hard to get it free from endogenous DNA. But other creatures must also contain RNA polymerase. It is an article of biological creed that all cells are alike under the double membrane, and in addition, just as we have been studying RNA polymerase in pea-embryo chromatin, other groups have been studying the enzyme in, not chromosomes, but micro-organisms. Thus Weiss of the University of Chicago has revealed the presence of RNA polymerase in *Microccocus*. Hurwitz and his allies at New York University simultaneously studied the enzyme in *Escherichia coli*. Berg and Chamberlain at Stanford also interested themselves in the RNA polymerase of *E. coli*. Severo Ochoa and his colleagues at New York University used *Azotobacter* as their subject for the study of this same enzyme, while Stephens at St. Louis University used *Salmonella* for the purpose. All of these groups have arrived at the same conclusion, namely, that creatures contain an RNA polymerase which conducts RNA synthesis using the riboside triphosphates as substrates and which conducts such synthesis only in the presence of DNA, the DNA serving as template for RNA synthesis. Let us consider further the RNA polymerase of *E. coli*. Packed cells of *E. coli*, grown in the appropriate manner contain per unit of fresh weight five hundred times as much RNA polymerase as pea embryos. Since cells of *E. coli* only contain one-thousandth as much DNA as do cells of peas, the amount of RNA polymerase per unit DNA in *E. coli* is five hundred thousand times greater than in peas. It is therefore no wonder that *E. coli* can produce messenger RNA and hence protein so much more rapidly than the pea plant.

Let us therefore prepare some RNA polymerase from *E. coli*. It will be useful because pea DNA supports the synthesis of RNA by such polymerase very well indeed (Fig. 1). Moreover, the native nucleohistone component of chromatin is inactive in the support of RNA synthesis by *coli* polymerase, although not itself inhibitory, since the polymerase of *E. coli* effectively uses pea DNA even in the presence of nucleohistone (Fig. 1).

It is desirable in this connection to analyze further this inactivity of native nucleohistone in RNA synthesis, so making a digression from our original digression. It is known that nucleohistones are exceedingly sensitive to magnesium ions and are in fact precipitated from solution by them in concentration as modest as 0·01 M. In the early days of nucleohistone biology, for example in early 1963, it was objected by various workers that the inactivity of native nucleohistone in support of RNA synthesis is due to its precipitation by the magnesium ions which are necessary in the reaction mixture in which RNA synthesis is tested. The precipitated-from-solution school, has, regrettably, enjoyed some vogue. It is however not correct since it can readily be demonstrated as follows that soluble

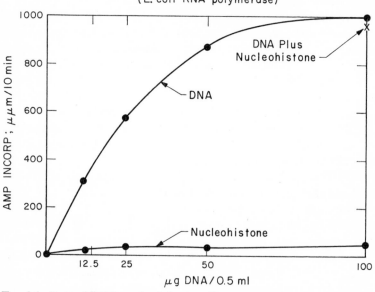

INEFFECTIVENESS OF CHROMOSOMAL NUCLEOHISTONE
IN SUPPORT OF RNA SYNTHESIS
(E. coli RNA polymerase)

FIG. 3.1. Rate of RNA synthesis by *E. coli* RNA polymerase as a function of DNA concentration. DNA added as deproteinized pea DNA or as native nucleohistone. X indicates 100 μg deproteinized DNA added in presence of 100 μg DNA as native nucleohistone (after Huang *et al.*, 1964).

nucleohistone remains soluble in the RNA synthesis reaction mixture. Soluble nucleohistone of sedimentation coefficient approximately thirty-two Sevdberg units was dissolved in the complete RNA synthesis reaction mixture, which includes ATP, UTP, GTP, and CTP, magnesium ions (0·004 M), manganese ions (0·001 M) beta-mercaptoethanol and tris buffer pH 8·0. The mixture was then centrifuged at one hundred and thirty times thousand times gravity which is sufficient to pellet any major aggregates but insufficient, in the twenty minutes employed, to sediment molecularly dispersed nucleohistone.

At the end of the centrifugation the contents of the tube were removed in successive portions and each assayed chemically for content of nucleohistone and biosynthetically after addition of RNA polymerase for ability to support RNA synthesis. The data of Table 1 show that nucleohistone remains in solution in the reaction mixture but is totally inert in support of RNA synthesis.

TABLE 1

Native pea nucleohistone is inactive in the support of RNA *synthesis even though it remains soluble in the incubation mixture used for support of such synthesis*

Fraction assayed	DNA μg/0·5 ml	RNA synthesized $\mu\mu$m/0·5 ml/10 min†
Top 1 ml‡	120	15
Second 1 ml	110	24
Third 1 ml	95	21
Bottom 1 ml	95	15
Average of above	105	19
Bottom 1 ml† DNA	195 (100 as free DNA)	1237

† Incorporation by enzyme alone (15 $\mu\mu$m) not subtracted.
‡ Soluble nucleohistone was centrifuged in complete standard mixture for RNA synthesis (except for RNA polymerase) for 20 minutes at 130000 × g. The contents of the tube were then removed in successive 1-ml portions. Each portion was assayed for DNA (as nucleohistone) content and for ability to support RNA synthesis after addition of polymerase.
After Huang *et al.*, 1964.

RNA polymerase of *E. coli* is then suitable for investigation of chromosomal matters, at least for matters of pea chromosomes. *Coli* polymerase can utilize pea DNA and cannot utilize the nucleohistone of pea chromatin, in both respects resembling pea chromosomal RNA polymerase. Let us therefore look into the purification of *E. coli* RNA polymerase, a procedure which has been studied in detail by Chamberlin and Berg (1962). First we must have some cells. Cells of *E. coli* vary greatly in polymerase content and in DNA/polymerase ratio, depending on cultural conditions and phase of growth. It is best to grow cells in a medium containing only mineral salts and glucose (1 per cent), vigorously aerated, and to harvest in middle log phase at a titer of 4×10^8 cells per ml. This is a low titer and correspondingly low yields of bacteria are obtained, less than one gram of packed cells per litre of culture. Such cells are however rich in polymerase and of low DNA/polymerase ratio and are essential to the preparation of enzyme of high dependency ratio. The cells may be harvested and stored at −20° for several months before use. We then merely grind some cells with glass beads in a blender, all at 5°C. Next we centrifuge off the glass beads and cellular debris at thirty thousand times gravity and then the ribosomes at one hundred and five thousand times gravity. The clear supernatant contains RNA polymerase, bound however to DNA. From now on, in our isolation of the enzyme, we must assay each step to determine if we are doing the correct thing. The assay consists merely in supplying some of the fraction with magnesium ions, beta-mercapto ethanol, the four riboside triphosphates (one of them C^{14}-labelled) and incubating the whole at 37°. One sample is incubated in the presence of and one in the absence of a saturating concentration of DNA. The reaction mixtures are then precipitated with trichloracetic acid (TCA) and the amount of TCA-insoluble RNA formed determined by counting of radioactivity. The use of this assay in the purification of RNA polymerase is greatly speeded by the use of millipore or similar filters, which permit passage of unused labelled riboside triphosphate, but retain RNA.

Our first step in the purification of RNA polymerase is the selective precipitation with streptomycin of DNA molecules which do not have RNA polymerase attached to them. This DNA is discarded. From the supernatant we next precipitate with protamine DNA molecules which do not have RNA polymerase attached. The precipitate, containing enzyme and DNA, is next treated with a solution containing a high concentration (0·1 M) of magnesium ions. This makes the DNA relatively insoluble and from the precipitate RNA polymerase may then be extracted with a suitable solvent (0·1 M ammonium sulphate) which selectively extracts enzyme from DNA. This extract we then fractionally precipitate with ammonium sulphate, discarding the fraction richest in nucleic acid. By this time we have lost little of the original enzyme but most of the DNA. The dependency ratio of the enzyme (ratio of activity in the presence and absence of added DNA) is now between 30 and 100 to 1. Further purification may be achieved by passing the enzyme through a DEAE column from which it emerges as a single peak. The dependency ratio now lies between 500 and ∞ to 1. RNA polymerase of *E. coli* prepared in this way may be stored in liquid nitrogen for several weeks without great loss of activity.

This then is the end of our digression. Let us return to chromatin. We have seen that chromatin heated to 60° to

TABLE 2

Activity of chromatin of developing pea cotyledons and of pea DNA in support of RNA synthesis by E. coli RNA polymerase

System†	RNA synthesis $\mu\mu$m nucleotide/10 min
125 μg DNA: *coli* polymerase	2220‡
37·5 μg DNA: *coli* polymerase	2030
12·5 μg DNA: *coli* polymerase	740
125 μg DNA as chromatin: *coli* polymerase	760
125 μg DNA as chromatin: no *coli* polymerase	17

† Reaction mixture includes tris, pH 8, 20 μm; $MnCl_2$, 0·5 μm; $MgCl_2$, 2 μm; 8-C^{14} ATP, 0·13 μm, 1·5 μc/μm; GTP, CTP, and UTP, each 0·2 μm; β-mercaptoethanol, 6 μm; chromatin or deproteinized DNA and *E. coli* polymerase (10 μg) as indicated, all in 0·5 ml final volume. Incubation at 37°.
‡ Incorporation by polymerase alone subtracted.
After Bonner *et al.*, 1964.

inactivate its endogenous polymerase, can serve as template in support of RNA synthesis by added, solubilized, chromosomal polymerase. Our *E. coli* polymerase serves equally well. In fact, since we have more of it in hand we can use more enzyme and make more RNA with a given amount of chromatin than with added chromosomal polymerase. Even more interesting and important, native, unheated chromatin can serve as template for the synthesis of large amounts of RNA by *E. coli* polymerase. Table 2 shows that the DNA of the chromatin of developing pea cotyledons is as effective in this function as about ten times less pure deproteinized DNA. The data of Table 2 also show that with *coli* polymerase the output of RNA by chromatin is actually increased above the level characteristic of chromatin which has available only its own endogenous polymerase. Rate of RNA synthesis by chromatin would appear to be limited by the amount of polymerase it contains. In the example of Table 2, rate of RNA synthesis is increased some fiftyfold by *coli* polymerase in the concentration used. But rate of RNA synthesis is linear with polymerase concentration in the concentration range used. If we use twice as much polymerase, rate of RNA synthesis by chromatin is increased some one hundredfold above the endogenous level.

We have in our hand, then, a new tool, a tool which serves not only as does the chromosomal polymerase as readout for the genetic information of the chromosome, but one which also and simultaneously serves as an amplifier of this information, making much messenger RNA available where little was available before. The supplementation, with *coli* polymerase, of the chromosomal system makes it possible to hear the chromosomal message loud and clear. Still, it must be a disconcerting experience for a molecule of *coli* RNA polymerase, used to seeing and working on DNA only in the form of the simple molecule present in *E. coli*, to suddenly find itself confronted for the first time with the vast structural framework of a chromosome.

Now to two further aspects of chromatin husbandry, aspects which we can study with exogenous RNA polymerase. We have developed in Chapter 2, two independent ways to measure the relative contributions to the genome of the nucleohistone and of the histone—not complexed components of chromatin. The first is based upon the fact that complexing with histone stabilizes DNA against melting. This causes chromatin to exhibit a two-step melting profile. The nucleo-nonhistone portion of chromatin melts with the T_M characteristic of pure DNA. The nucleohistone component melts only at a higher temperature and with the T_M characteristic of nucleohistone. From the relative contributions of the two steps we can deduce the relative contributions of the two forms of DNA to the genome. A second way of getting at this same matter is by mechanical shearing of chromatin. By this procedure we recover the nucleohistone component as soluble material and the nucleo non-histone as an aggregated pellet. The distribution of the chromosomal DNA between these two fractions shows the contribution of each to the genome. Now we can devise a third technique to approach the same question. We can determine the effectiveness relative to pure DNA of a given amount of DNA in the form of chromatin, in support of RNA synthesis in the presence of excess RNA polymerase. How well do these three methods agree with one another? Such a comparison is made in Table 3. The data concern chromatin of three different origins. The first is from pea embryos, the

TABLE 3

Properties of various chromatins

Property	Pea-embryo chromatin	Chromatin of developing cotyledons	Chromatin of duck erythrocytes
Fraction of total DNA which melts with $T_M = 70°$	20%	about 5%	about 0%†
Fraction of total DNA recovered in nucleohistone	78%	about 93%	98%
Fraction of total DNA recovered in nucleo non-histone	22%	about 7%	about 0–2%
Effectiveness relative to pure DNA in support of RNA synthesis by polymerase	about 20%	about 10%	0·1%

† DNA of duck, like that of pea and thymus, exhibits a T_M of 70° in dilute saline citrate (0·016 M).

second is from developing pea cotyledons, and the third is from the nucleated erythrocytes of the duck. The data of Table 3 indicate how widely the proportion of low melting and hence nucleo non-histone material varies from one kind of chromatin to another. In the pea embryo this fraction is some 20 per cent of the total, in developing pea-cotyledon chromatin about 5 per cent, and in the chromatin of duck erythrocytes essentially nothing. The mechanical fractionation procedure reveals similar differences in the contributions of nucleohistone and nucleo non-histone to chromatin composition. Thus about 22 per cent of the DNA of pea-embryo chromatin is recovered in the nucleo non-histone fraction but only about 7 per cent of the DNA of cotyledon chromatin, while duck erythrocyte chromatin yields essentially none. The findings obtained by these two quite different methods certainly agree with one another satisfactorily.

Now let us determine for each chromatin the proportion of its DNA which is derepressed, which is active in supporting DNA-dependent RNA synthesis in the presence of added *E. coli* RNA polymerase. The data, included in Table 3, show again that embryo chromatin is effective in support of RNA synthesis. Duck erythrocyte chromatin is almost totally ineffective while cotyledon chromatin is intermediate. The results of this test then agree with those obtained by the two physical measurements. We can begin to feel with some confidence that the various measures of chromosome structure and composition do possess some biological meaning. It appears to be the low melting portion of the DNA of chromatin which supports chromosomal RNA synthesis.

We have been concerned with the use of polymerase of *E. coli* to transcribe the chromosomal message of pea plants. Can we generalize from the pea plant to other creatures? The answer is yes. Table 4 shows that the methods found successful for preparing pea-plant chromatin are also successful for preparing chromatin from a variety of organs of a variety of plants and animals. And all of these kinds of chromatin will support, in the presence of added *E. coli* RNA polymerase, the synthesis of much more RNA than is supported with their endogenous chromsomal RNA polymerase alone. The parallel between the behaviour of pea-plant chromatin and the chromatins of other organs goes still further. Two-step melting, fractionation by shearing into two components, inactivity of the nucleohistone component in support of RNA synthesis— these properties are common to many kinds of chromatin with the exception of those chromatins like that of duck erythrocytes which appear to be completely repressed.

The ability of chromatin to support DNA-dependent RNA synthesis in the presence of *E. coli* RNA polymerase, can handily be used to detect the presence of chromatin in a homogenate or cell fraction and can be used as a guide during its purification. The method is fast, and materials which interfere with spectrophotometric or diphenylamine determinations of DNA do not interfere with it. It is also highly sensitive. We have used it as

TABLE 4

Chromatin† of varied organs of varied creatures. All of the kinds of chromatin listed support RNA synthesis in the presence of E. coli RNA polymerase

Creature	Organ
Pea	Embryo
	Cotyledon
	Root
	Stem
	Apical bud
Rat	Liver
	Kidney
	Spleen
Duck	Liver
	Reticulocyte
Potato	Bud (non-dormant)

† Chromatin in each case isolated by methods similar to those used for peas.

a guide to the isolation of chromatin from a tissue in which large amounts of particulate protein makes spectroscopy impossible and in which gums make other analytical procedures useless. Ability to support DNA dependent RNA synthesis is a basic property of chromatin.

We have then a general operational chromosomal husbandry, and a general, if partial, theory of chromosome structure. We have also a general method not only for the transcription but also for the amplification of the chromosomal message. We are now nearly ready to apply our new-found lore to our original task of finding out how genetic activity is controlled in the chromosome.

SELECTED REFERENCES

E. coli RNA *polymerase*

CHAMBERLIN, M. and BERG, P., *Proc. Nat. Acad. Sci., Wash.* **48**, 81 (1962).

Nucleohistone and microbial RNA *polymerase*

HUANG, R. C. C., BONNER, J. and MURRAY, K., *J. mol. Biol.* **8,** 59 (1964).

Fortification of chromatin with E. coli RNA *polymerase*

HUANG, R. C. C., BONNER, J. and GILDEN, R., *Proc. Nat. Acad. Sci., Wash.* **50,** 893 (1963).

TUAN, D. Y. and BONNER, J., *Plant Physiol.* **39,** 768 (1964).

MARUSHIGE, K. and BONNER, J., *Fed. Proc.* **23,** 165 (1964).

4

THE TRANSCRIPTION OF DNA

THE concept of transcription of DNA into RNA was first forcibly brought to the attention of biologists by a classic experiment of Volkin and Astrachan. In this experiment, the DNA of bacteriophage was allowed to penetrate into bacterial cells which were also supplied, after a short time lag to ensure RNA synthesis by the host had ceased, with P^{32}-labelled orthophosphate. It was shown that the viral DNA results in the formation in the host cell of RNA which possesses a remarkable property, namely, a base composition similar to that of the viral DNA upon which its synthesis is dependent. In the viral DNA, $dA = dT$ and $dG = dC$. The newly formed RNA was also found to have base complementarity with $A = U$ and $G = C$.

The experiment of Volkin and Astrachan provided early support for today's central dogma—namely that the DNA makes the RNA which makes the protein. As a result, the feeling grew among biologists that messenger RNA made by transcription of DNA, must mirror the base composition of its DNA template. This feeling was strengthened by base analyses of the RNA formed *in vitro* by transcription of DNA by RNA polymerase. The extensive studies of Chamberlain and Berg (1962) and of Hurwitz and his group (1962) have shown clearly and elegantly that the RNA synthesized has both the base composition and the nearest-neighbour frequencies of its template DNA. Thus there developed the view, almost an article of faith, that biologically important and interesting RNA must exhibit base complementarity and a base composition similar to that of its parental DNA. These criteria have even been used to assess whether or not a particular nuclear RNA is, in fact, messenger RNA. (Allfrey and Mirsky, 1962: Allfrey *et al.*, 1963: Sibitani *et al.*, 1962.)

Today more is known about the transcription of DNA by RNA polymerase, both in living cells and in the test tube.

It is now clear that the RNA made in living cells does not in general possess the composition of its template DNA, and that this is because, in life, but a single strand of the double-stranded DNA is transcribed by RNA polymerase. The cell possesses mechanisms which determine which strand is transcribed. We will, in this chapter, summarize the facts concerning the transcription of DNA by RNA polymerase.

The first relevant fact is that in the transcription of DNA by RNA polymerase in the test tube, it is a single strand of DNA which is read. Thus single-stranded DNA serves effectively as a template for RNA synthesis *in vitro*. The single strand may be one made by denaturing native DNA or it may be one which occurs naturally as does the single-stranded DNA φX-174 virus. φX DNA is not base complementary. The RNA made using φX DNA as template is not base complementary to itself but is base complementary to φX DNA.

It is then clear that the transcription from double stranded DNA of an RNA possessing the same base composition requires, in general, that both strands be transcribed. In a single strand of a double helical DNA molecule, dA need not equal dT nor dG equal dC. It is only in the double helical molecule that there is such equivalence. In the case of the RNA synthesized by DNA-dependent RNA synthesis, the RNA made when a single-stranded DNA chain is used as template will be base complementary to the template (as in the case of φX-174 supported RNA synthesis) but will not in general be base complementary to itself. RNA synthesized *in vitro*, with heat-denatured DNA (a mixture of the two separated complementary DNA strands) as template, has the base composition of the DNA and is base complementary to itself. Such RNA must consist of a mixture of equal numbers of two complementary species of RNA chains formed by transcription of the two complementary DNA chains. How then do we imagine the *in vitro* transcription of native double-stranded DNA? Clearly, both strands must be transcribed, forming a mixture of equal numbers of two complementary species of RNA chains. We imagine, although we do not yet know with certainty, that RNA polymerase attacks an end or other designated port of entry of the DNA molecule and marches down the chain of proper polarity separating the two strands of DNA as it goes, leaving behind a growing chain of RNA complementary to the chosen DNA strand. The growing RNA strand will initially be bound to its template DNA in DNA-RNA hybrid form. The DNA-RNA is, however, less stable than the corresponding DNA-DNA duplex. The second free (non-template) DNA strand will then continuously displace the newly synthesized RNA strand from the template DNA with reformation of the original DNA duplex (Paigen, 1962).

It is then clear that when native DNA is transcribed by RNA polymerase *in vitro*, both strands are in fact transcribed to yield RNA of base composition complementary to itself and identical with that of the template DNA. We know, however, that this cannot happen in life. For example, the transfer and ribosomal RNA's of peas and of *E. coli* are not base complementary to themselves and they are made by transcription of DNA (Chapter 7). The same is true of messenger RNA produced by transcription of pea chromatin *in vitro*. Even the messenger RNA's of phage T2 and of phage T4 are not really base complementary to themselves as reinvestigation of the phenomenon studied by Volkin and Astrachan has shown (Bautz and Hall, 1962). Can it be then that in life but a single strand of DNA is transcribed? That this is in fact so has been clearly and elegantly demonstrated by Hayashi *et al.* (1963). They used φX-174 DNA which, in the vegetative phage, is single-stranded and not base complementary. Upon its entry into the host cell, the complement to the vegetative strand is formed resulting in a duplex—the replicative form of φX-174 (Sinsheimer *et al.*, 1962). Only after synthesis of the replicative form does synthesis of φX-174 RNA commence. This phage-specific messenger RNA then accumulates and reaches substantial proportions by fifty minutes after infection. By the use of this information, it was isolated, free from host RNA and in tritium-labelled form by column chromatography (Hayashi *et al.*, 1963). The data of Table 1 show that this messenger RNA can form hybrid RNA-DNA double helices, can hybridize (considered in detail in Chapter 7) only with denatured replicative DNA, and cannot hybridize with the single-stranded vegetative DNA. This shows that the messenger RNA of the phage is complementary to but one of the two DNA strands of the replicative form—namely to that strand which is formed

TABLE 1

Messenger RNA *produced in the host cell as the result of infection with bacteriophage* φX-174 *is capable of forming* DNA-RNA *hybrid double helix with denatured* DNA *derived from double-stranded helical* φX DNA (replicative form) *but not with single-stranded vegetative* φX DNA

φX virus-dependent messenger RNA allowed to form hybrid with DNA derived from:	Amount of hybrid RNA-DNA helix formed (cpm)
Denatured φX replicative form (double-stranded)	3473
Vegatative φX DNA (single-stranded)	<60

After Hayashi *et al.*, 1963.

after infection and which is complementary to the vegetative strand. This conclusion is supported by analyses of the base composition of the messenger RNA as shown in Table 2. The phage RNA is identical in base composition to the vegetative DNA strand of φX-174 and hence complementary to the strand peculiar to the replicative form.

Similar results have been obtained for the messenger RNA produced after infection of bacterial cells by two other bacteriophages. The infection of *Bacillus megaterium* by phage α has

TABLE 2

Base composition of φX-*dependent messenger* RNA *and of the two strands of* φX DNA

Nucleic acid	Base composition: moles base/100 moles			
	C	A	U(T)	G
φX-M RNA	17·5	25·5	34	23
φX-DNA, vegetative strand	19	25	33	23
φX-DNA, complimentary	23	33	25	19

After Hayashi *et al.*, 1963.

been studied by Toccini *et al.* (1963). The DNA of phage α is double-stranded but the two strands have base compositions and therefore densities sufficiently different for them to be separatable by centrifugation which yields a light and a heavy α-DNA. The phage RNA synthesized in the host cell was pulse-labelled and isolated. This RNA hybridized with only one of the two strands of the α-DNA—namely with the heavy strand. The base composition of the hybridized RNA is complementary to that of the heavy strand of the α-DNA and identical with that of the light strand.

Similar to these findings are those of Marmur and Greenstamm (1963) on the phage SP 8 of *Bacillus subtilis*. The two strands of the DNA of this virus are also sufficiently different in density so that they can be physically separated by centrifugation on a cesium chloride gradient. Here again messenger RNA of the phage, isolated from the infected host, hybridizes with but one of the two strands of DNA.

Whenever purified DNA of these same phages is transcribed by RNA polymerase *in vitro* the RNA formed is base complementary to itself and hence is a transcription of both strands of the template DNA. In life (and with isolated chromatin), but one strand is transcribed. Quite evidently, the genomal DNA, as it exists in life, or in isolated chromatin, is different from purified DNA—it contains a control mechanism which determines that only one particular strand of DNA is transcribed. What is this control mechanism and what purpose does it serve?

We know, of course, that when we prepare pure DNA from chromatin, we remove histone from the nucleohistone component. But we also extensively degrade the chromosomal DNA into pieces shorter than the original. We produce new DNA ends. The same thing occurs in the purification of the DNA of phage and even of bacteria. Let us look at these simpler cases first.

DNA prepared from *E. coli* by the classical and gentle procedure of Marmur (1961) has a molecular weight of approximately 10×10^6. This is a high molecular weight yet we are now certain that the DNA has been degraded by mechanical shear during isolation. Cairns (1963) has prepared DNA from *E. coli* by a method in which shear is minimized; the cell membrane is ruptured and the DNA allowed to leak slowly

4

out of the cell. Electron microscopy shows that the unsheared DNA of *E. coli* is present as a single particle of molecular weight 2000×10^6. Extensive shearing must therefore occur in the gentlest of procedures for bulk preparation of DNA. Cairns has further shown that the *coli* DNA forms a circle, a ring, and is thus endless. This structure is, of course, lost during the bulk preparation of *coli* DNA.

The characterization of the bacterial genome—namely that all of the genomal DNA is contained in a single ring-shaped molecule have now been found also in several viruses. That the intact genome of the polyoma virus is a DNA ring of molecular weight approximately $2 \cdot 8 \times 10^6$ has been shown by Weil and Vinograd (1963). The ϕX-174 genome is itself a ring molecule and genetics indicate the same to be true of phage T4. In all of these cases, the rings are, however, broken by chain scission during DNA isolation unless particular care is taken. The purified DNA of pea chromatin is, as we have seen, composed of molecules of molecular weight approximately 9×10^6. These too, appear however, to be artifacts due to shearing. Hyde, in our laboratory, has prepared chromosomal DNA by methods analogous to those used by Cairns, which minimize shearing. Isolated intact nuclei are merely placed upon the surface of solution of high ionic strength ammonium acetate. The chromosomal DNA as it leaks out is complexed with a positively charged protein to make possible visualization of the DNA by electronmicroscopy. It is not readily possible to completely abolish all shear forces in this procedure. Such forces are apparently introduced by the very act of disruption of the nucleohistone complex. Nonetheless it is clear from Hyde's work that the DNA strands in native chromatin are very much longer than those in purified chromosomal DNA.

We see then that the procedures by which DNA is isolated and purified are procedures which introduce new strand ends—ends which are not present in the genome as it exists in nature. It is possibly this fact that enables the RNA polymerase to indiscriminately read both strands *in vitro* starting as it can from either end of the double-stranded molecule. But in life a much smaller number of strand ends appear to be available for this purpose and, in fact, in the ring DNA of bacterial viruses one strand would appear to have no end at all. Possibly this situation may also well be true with the DNA of chromosomes.

This is, however, still not the whole story. A further method of control of transcription is available in life, a method for which we as yet have no model. That this is so is clear from the fact that ring-form vegetative ϕX-174 DNA, which is not transcribed in life, is readily transcribed *in vitro* by RNA polymerase. The resulting duplex is then a ring-form RNA-DNA hybrid. Clearly, the discovery of the way in which transcription of a single chain of a DNA duplex is governed in life is a central and most important matter.

The replication of DNA by DNA polymerase is somewhat different in properties from transcription. Cairns (1963) has shown, by radioautography, that the chromosome of *E. coli* grows at but a single point. Thus, a point of DNA replication at which both chains of the DNA are doubled, moves around the ring chromosome. The same has also been shown in different ways by Nigata (1963), by Yashikawa and Sueoka (1963), and by Bonhoeffer and Gierer (1963).

The question of the number of DNA growing points involved in DNA replication is, of course, a suitable kind of question to ask about chromosomes. An answer to the question would help to settle how many continuous DNA molecules there are in a chromosome. This has been studied in a Cairns-type of experiment with *Drosophila* salivary gland chromosomes by Plaut (1963). The result is clear. Each band of the polytene chromosome contains a DNA growing point (actually, growing points presumably equal in number of the polyteny of the chromosome). This result favours the picture of the chromosome as a succession of individual long molecules rather than as a single very long molecule.

How long are these pieces? In the pea plant each chromatid of each chromosome has a length corresponding to a particle weight of 250×10^9 (each chromatid of each chromosome is therefore about 10^9 nucleotides long). These chromatids contain therefore an amount of DNA equal to that contained in somewhat over 100 *E. coli* chromosomes. It would not be surprising if pea chromosome were made up of 100 replicating units—each of the size of the *coli* chromosome. They would then be of the order suggested by the experiment of Plaut with *Drosophila*.

In summary, we have seen that in life or in the isolated chromosome, but a single strand of the double helical DNA is transcribed. The product RNA consists, therefore, of bachelor strands—strands which have no mate with whom to hold hands and to form double helical structures. The single strand transcription of DNA is, in fact, an important feature of the strategy of life. It assures that messenger RNA, by being born single-stranded, will be able to base pair with molecules of transfer RNA—also born single-stranded. It is upon this recognition by base pairing of messenger RNA codons by transfer RNA anti-codons, that the sequencing of amino acids in enzyme molecules is founded. We have also seen too that the machinery which controls the transcription by RNA polymerase of but a single strand of the genomal DNA, is preserved in isolated chromatin but is not preserved in isolated DNA. Understanding of how transcription is confined to a single strand will no doubt deepen our understanding of how genetic activity is controlled.

SELECTED REFERENCES

Early work on relation of RNA *composition to* DNA

VOLKIN, E. and ASTRACHAN, L., *Virology* **2**, 149 (1956).

CHAMBERLIN, M. and BERG, P., *Proc. Nat. Acad. Sci., Wash.* **48**, 81 (1962).

FURTH, J. J., HURWITZ, J. and ANDERS, M., *J. Biol. Chem.* **237**, 2611 (1962).

HURWITZ, J., FURTH, J., ANDERS, M. and EVANS, A., *J. Biol. Chem.* **237**, 3752 (1962).

Some views extrapolated from the above

ALLFREY, V. G. and MIRSKY, A. E., *Proc. Nat. Acad. Sci., Wash.* **48**, 1590 (1962).

ALLFREY, V. G., LITTAU, V. C. and MIRSKY, A. E., *Proc. Nat. Acad. Sci., Wash.* **49**, 414 (1963).

SIBITANI, A., de KLOET, S. R., ALLFREY, V. G. and MIRSKY, A. E., *Proc. Nat. Acad. Sci., Wash.* **48**, 471 (1962).

The physical nature of transcription

GAJDUSEK, E. P., NAKAMOTO, T. and WEISS, S. B., *Proc. Nat. Acad. Sci., Wash.* **47**, 1405 (1961).

CHAMBERLIN, M., BALDWIN, R. and BERG, P., *J. mol. Biol.* **7,** 334 (1963).

WARNER, R. C., SAMUELS, H. H., ABBOT, M. and KRAKOW, J. S., *Proc. Nat. Acad. Sci., Wash.* **49,** 533 (1963).

SINSHEIMER, R. L. and LAWRENCE, M., *J. mol. Biol.* **8,** 289 (1964).

CHAMBERLIN, M. and BERG, P., *J. mol. Biol.* **8,** 297 (1964).

Transcription from a single DNA strand of a double-stranded DNA

BAUTZ, E. and HALL, B., *Proc. Nat. Acad. Sci., Wash.* **48,** 480 (1962).

HAYASHI, M., HAYASHI, M. N. and SPIEGELMAN, S., *Proc. Nat. Acad. Sci., Wash.* **50,** 664 (1963).

TOCCHINI, G., STODOLSKY, M., AURISICCHIO, A., SARNAT, M., GRAZIOSI, F., WEISS, S. B. and GAJDUSEK, E. P., *Proc. Nat. Acad. Sci., Wash.* **50,** 935 (1963).

MARMUR, J. and GREENSPAN, C., *Science* **142,** 387 (1963).

Basic work with φX-174, a single-stranded DNA virus

SINSHEIMER, R., *J. mol. Biol.* **1,** 43 (1959).

SINSHEIMER, R., STARMAN, B., NAGLAR, C. and GUTHRIE, S., *J. mol. Biol.* **4,** 142 (1962).

Length and nature of DNA

MARMUR, J., *J. mol. Biol.* **3,** 208 (1961).

CAIRNS, J., *J. mol. Biol.* **6,** 208 (1963).

WEIL, R. and VINOGRAD, J., *Proc. Nat. Acad. Sci., Wash.* **50,** 730 (1963).

Number of DNA growing points

CAIRNS, J., *J. mol. Biol.* **6,** 208 (1963).

NAGATA, T., *Proc. Nat. Acad. Sci., Wash.* **49,** 551 (1963).

YOSHIKAWA, H. and SUEOKA, N., *Proc. Nat. Acad. Sci., Wash.* **49,** 559 (1963).

BONHOEFFER, F. and GIERER, A., *J. mol. Biol.* **5,** 534 (1963).

PLAUT, W., *J. mol. Biol.* **7,** 632 (1963).

5

THE DECODING OF MESSENGER RNA

THE DNA dependent synthesis of RNA by chromatin can be increased by two orders of magnitude by the addition of exogenous RNA polymerase, and this polymerase may be derived from a different creature than the chromatin. Since it is possible in this way to remove uncertainties concerning the distribution and availability of chromosomal RNA polymerase within the genome, it is possible to test the extent to which the genomal DNA of chromatin is active in the support of RNA synthesis. We have seen that most chromatins support RNA synthesis less well than an equal amount of pure DNA. Chromatin behaves as though only a portion of its DNA is active.

We may now return to the subject of Chapter 1, namely, that of the decoding of the messages contained in the RNA synthesized under chromosomal auspices. To do this we couple our chromosomal RNA generating system to a system which synthesizes protein. The protein-synthesizing strategy of nature is based upon the ribosome. We borrow the ribosomes of *E. coli* for simple reasons of greediness; they are the best. Many people have contributed to working out in detail optimal preparative procedures for *coli* ribosomes. As a result, isolated *coli* ribosomes stand above all others in rate of protein synthesis, in dependence upon added messenger RNA, and in release of finished protein. A fact which could not have been predicted in advance, is that *coli* ribosomes are fully competent to read RNA generated from chromosomes of higher creatures, at least of peas. The messages are different in different cells and in different creatures, but ribosomes themselves appear to be all much the same.

Let us summarize the principles of ribosomal preparation. The principle of particular importance to us is that we have a ribosomal system highly dependent upon added messenger

RNA—a system free of endogenous messenger. We first grow some *E. coli*. We have seen that the best *coli* cells for preparing RNA polymerase are harvested in middle log phase. This is not true of ribosomal preparations and the best ribosomes are made from *coli* harvested in late log phase, at concentrations of 8 to 10 $\times$ 10^8 cells per ml giving two grammes fresh weight of packed cells per litre of culture medium. The frozen cells are placed in an old-fashioned mortar pre-chilled to $-80°$ and ground to a powder with a large excess of alumina powder. The alumina grains must be of appropriate size, slightly larger than the bacteria; grains of five microns do well. The powder is quickly transformed into a slurry with buffer and centrifuged at 30000 $\times$ g to remove alumina powder and cell fragments. From the supernatant, ribosomes may be pelleted at 105000 $\times$ gravity, resuspended and repelleted and taken up finally in small volume. Since a ribosomal system requires soluble enzymes such as those which activate amino acids as well as transfer RNA's and which are not pelleted at 105000 $\times$ gravity it is necessary to add some of the supernatant from the 105000 $\times$ g pelleting of the ribosomes. The ribosomal system for protein synthesis is also fortified with the twenty amino acids of which proteins are made, ATP as a source of energy for the formation of peptide bonds, and GTP which is required for transfer of amino acid from acyl-S-RNA complex to ribosome. Suppose we now incubate our system for say, 30 minutes, and see whether protein has been formed. It has. This is because the ribosomal system prepared in this standard way yields ribosomes and supernatant which contain messenger RNA. Protein synthesis by such conventionally prepared ribosomal systems depends on residual, surviving messenger RNA. Such a ribosomal system is of course, unsuitable for our present work. We must have our system free, or as free as possible, of messenger RNA and of the ability to make it. Our ribosomal system may be freed of messenger RNA by a procedure suggested by Nirenberg and Matthaei (1961) and by Ning and Stephens (1962). We go back to the supernatant from the first centrifugation at 30000 $\times$ gravity. To it we add all of the substrates required for protein synthesis, the twenty amino acids, ATP etc., and incubate for 45 to 60 minutes at 37°. Under these circumstances the ribosomes make protein, but, in addition, the messenger RNA is degraded. The half-life of messenger RNA in a living bacterial cell is of the order of 1 to 2 minutes. Since no riboside triphosphate other than ATP is provided, the synthesis of new messenger RNA does not take place and the system becomes depleted of it. The pre-incubated homogenate can now be centrifuged at 105000 × g to obtain the ribosomes, which are then resuspended, repelleted, resuspended, and exhaustively dialyzed to free them of amino acids etc. The supernatant from the initial pelleting of ribosomes is also exhaustively dialyzed to free it of amino acids and provides the amino acid activating enzymes and transfer RNA. Unfortunately, a ribosomal system prepared in this way still synthesizes protein if it is incubated with the complete mixture of all amino acids, and the four riboside triphosphates. This is because it contains *E. coli* DNA as well as some *coli* RNA polymerase and can therefore generate its own messenger RNA if it is provided with the riboside triphosphates. We must free our system of these materials. One method is precipitation of DNA and polymerase with protamine as we did earlier in preparing *E. coli* RNA polymerase. Protamine precipitates not only bacterial DNA but also transfer RNA. Hence after protamine precipitation we must also prepare pure *E. coli* transfer RNA and add it back to the supernatant. We have found it simpler to remove bacterial DNA together with its associated polymerase by centrifugation at 105000 × gravity for 24 hours. The bacterial DNA is quantitatively pelleted, and the supernatant is freed or nearly so of both bacterial DNA and RNA polymerase. We now have a ribosomal system which is dependent on messenger RNA. Table 1 concerns the properties and dependency aspects of such a system. The complete system contains ribosomes, proteins of the 105000 × g supernatant, transfer RNA, the twenty amino acids, the four riboside triphosphates, DNA of whatever kind it is desired to test, and RNA polymerase. The data of Table 1 show that our complete system is able, in the presence of pea DNA, to incorporate over one millimicromole of leucine into protein per milligram ribosomal protein per thirty minutes. Elimination from the system of either DNA, polymerase, or both, reduces protein synthesis by a factor of 10 or more. Our system possesses

TABLE 1

Dependence of protein synthesis by E. coli ribosomal system on pea DNA and E. coli RNA polymerase

Additions to ribosomal system†	Leucine incorp. into protein $\mu\mu$m/mg ribosomal protein/30 min
Pea DNA; RNA polymerase	1158
T2 DNA; polymerase	828
T4 DNA; polymerase	756
Polymerase: no DNA	100
Pea DNA; no polymerase	102
No DNA; no polymerase	104

† Reaction mixture includes tris, pH 8, 50 μm; MgAc$_2$, 2 μm; MnCl$_2$, 0·5 μm; KCL, 20 μm; β-mercaptoethanol, 3 μm; 18 amino acids, each 0·02 μm; C^{14}-leucine, 0·075 μm, 6·8 μc/μm; ATP(K), 1 μm; GTP, CTP, UTP, each 0·1 μm; 50 μg ribosomal protein, 69 μg; 105 kg supernatant protein, DNA (100 μg), and *E. coli* RNA polymerase (20 μg) as indicated, all in total volume of 0·3 ml. Incubation at 37°.
After Bonner *et al.*, 1963.

therefore a dependency upon added polymerase and DNA of over tenfold. The data of Table 1 show further that pea DNA is as effective in supporting ribosomal protein synthesis as is DNA of bacteriophages T2 or T4, DNAs which have been widely used in the study of both DNA dependent RNA synthesis and DNA dependent protein synthesis.

The data of Table 2 illustrate the added dependency conferred upon the ribosomal system by the pre-incubation outline above. Dependency upon the messenger RNA generating system is only about threefold if the system is prepared from non pre-incubated *coli* homogenates. Pre-incubation increases this dependency to tenfold; it is a worthwhile step.

Finally we may, if we wish, replace the DNA-RNA polymerase messenger RNA generating system by a chromatin-polymerase messenger RNA generating system. In the experiment of Table 3, chromatin and added *coli* RNA polymerase supplies the messenger RNA for the ribosomal system. Protein is synthesized. In the absence of added polymerase, protein synthesis is very much diminished, although still about twice as great as when

TABLE 2

Dependency of ribosomal protein synthesis on exogenous messenger RNA as affected by pre-incubation of ribosomal system in complete protein synthesis reaction mixture

System†	C^{14}-leucine incorporated into protein $\mu\mu m/60$ min/mg ribosomal protein	
	Pre-incubated	Not pre-incubated
Ribosomal system alone	88	251
Ribosomal system + pea DNA + RNA polymerase	844	754

† Reaction mixture contains tris pH 8·0, 50 μm; MgAc2, 2 μm; MnCl2, 0·5 μm; KCl, 20 μm; β-mercapto ethanol, 3 μm; 18 amino acids, each 0·02 μm; C^{14}-leucine, 6·8 μc/μm, 0·075 μm; ATP(K), 1 μm; GTP, CTP, UTP, each 0·1 μm, 50 μg ribosomal protein; 69 μg 105000 × g supernatant protein; pea DNA, 100 μg; and approx. 20 μg *E. coli* RNA polymerase, all in a total volume of 0·3 ml. Incubation at 37°.
After Bonner *et al.*, 1963.

TABLE 3

Support of ribosomal protein synthesis by chromatin-dependent RNA synthesis

System†	C^{14}-leucine incorporated into protein (cpm/30 min)	
	One-step expt.	Two-step expt.
Cotyledon chromatin: RNA polymerase: ribosomal system	721	505
Cotyledon chromatin: ribosomal system: to polymerase	133	87
Bud chromatin: RNA polymerase ribosomal system	924	182
Bud chromatin: ribosomal system: no polymerase	110	122

† The reaction mixture contains all materials required for both RNA and protein synthesis. In the two-step experiment, RNA is generated by chromatin for 30 min, the chromatin then removed by centrifugation and the ribosomal system then added to the supernatant. In the one-step experiment, all ingredients are present simultaneously.
After Bonner *et al.*, 1963.

both chromatin and polymerase are eliminated. Chromatin does of course possess some ability to synthesize RNA using its own chromosome polymerase. The data of Table 3 also concern two different strategies for the conduct of chromosomally-dependent protein synthesis. By one strategy, chromatin, polymerase, and ribosomal system are mixed simultaneously. Messenger RNA is formed and at once used by the ribosomal system. By the second strategy, chromatin, RNA polymerase,

TABLE 4

Release from ribosomes of completed, soluble protein, synthesized in response to messenger RNA made by pea cotyledon chromatin dependent RNA synthesis.†

Fraction‡	C^{14}-leucine incorporated into protein. $\mu\mu$m/mg ribosomal protein/30 min.
Total incorporated into protein	736 $\mu\mu$m
Total soluble protein	587 $\mu\mu$m
Per cent of total released as soluble protein	80%

† Reaction mixture contains all materials required for both RNA and protein synthesis as specified in Table 2. 50 μg DNA per 0·3 ml. reaction mixture supplied as chromatin of developing pea cotyledons. Incubation for 30 minutes at 37°.

‡ For determination of total incorporation into protein, an aliquot of the reaction mixture was precipitated and washed with 5% TCA at the end of the incubation period. For the determination of total soluble protein, a portion of the reaction mixture was, after incubation, centrifuged for 2 hours at 105,000 × g to remove chromatin and ribosomes. The supernatant proteins were then precipitated and washed with 5% TCA.

and the four riboside triphosphates are incubated together for 30 minutes. RNA is formed. The reaction mixture is then cooled, and the chromatin centrifuged off, the RNA formed remaining in solution. To their solution, containing chromosomally generated RNA, the ribosomal system is added and incubation continued for a further 30 minutes. Protein is formed. It is clear from Table 3 that such protein synthesis depends on the presence in the initial reaction mixture of RNA polymerase. This experiment shows clearly that chromatin generates RNA which is then able to support protein synthesis by the ribosomal system.

It is of some interest to consider the quantitative aspects of the support of ribosomal protein synthesis by messenger RNA. We have found the optimum reaction mixture for ribosomal protein synthesis to consist, of, per 50 micrograms of ribosomal protein, 70 micrograms of supernatant soluble enzymes supported by 20 millimicromoles of freshly generated RNA. This roughly corresponds to one messenger RNA molecule 1000 nucleotides long to each ribosome.

Although many *in vitro* ribosomal systems readily incorporate amino acids into peptide linkages, still, as has been repeatedly emphasized in the literature, little of this freshly made protein is ordinarily liberated from the ribosome as finished soluble product. Thus ribosomes of mamallian reticulocyte whose synthesis of haemoglobin *in vitro* has been much studied, release as soluble globin half or less of the protein they synthesize. Ribosomal systems from liver liberate essentially none of the amino acids which they incorporate into protein. It is noteworthy that this is not found with our complete protein synthesizing system; a ribosomal system which is supported not by residual messenger RNA but by messenger RNA which is continuously generated. Not only is synthesis of protein much more extensive than with ribosomal systems which do not contain a messenger RNA generator, but in addition the great bulk of the protein formed is liberated as soluble finished product. As much as 80 per cent of the newly formed protein remains in solution after the chromatin and ribosomes are removed by centrifugation. We conclude that the liberation of finished protein by the ribosomal system depends on the messenger RNA being intact. In ribosomal systems supported by residual messenger RNA it is probable that this RNA is fragmented. It is likely that the messenger RNA molecule contains an initial message unit which shows the ribosome where to start, and a final message unit which says that the protein has been synthesized and that it is time to release it. If this final message unit is lost, if the messenger RNA molecule is not intact, complete protein molecules are not synthesized and released. These are at least the conclusions to be drawn from our understanding of ribosome—messenger RNA interaction, work summarized in the next chapter.

SELECTED REFERENCES

Messenger RNA-*dependent ribosomal systems (from bacteria)*

NIRENBERG, M. and MATTHAEI, H., *Proc. Nat. Acad. Sci., Wash.* **47,** 1588 (1961).

WOOD, W. and BERG, P., *Proc. Nat. Acad. Sci., Wash.* **48,** 94 (1962).

NING, C. and STEPHENS, A., *J. mol. Biol.* **5,** 650 (1962).

FURTH, I., KAHAN, F. and HURWITZ, J., *Biochem. biophys. Res. Commun.* **9,** 337 (1962).

Chromosomally supported messenger dependent ribosomal systems

BONNER, J., HUANG, R. C. and GILDEN, R., *Proc. Nat. Acad. Sci., Wash.* **50,** 893 (1963).

6

THE INTERACTION OF RIBOSOME AND MESSENGER

To ROUND out our discussion of the readout of messenger RNA by ribosome, let us further consider our understanding of the physical interaction of these two entities. This understanding may be said to have begun with the demonstration by Nirenberg and Matthaei (1962) that in the presence of a particular and quite unusual type of messenger RNA—namely poly U—the ribosome produces an unusual polypeptide, namely polyphenyl alanine. This led in turn to the finding first indicated by Spyrides and Lipmann (1962) that in the poly U-supported synthesis of polyphenyl alanine, the ribosome not only binds to the poly U, but that, in addition, many ribosomes bind to a single poly U chain. Only those ribosomes bound in clusters to a single poly U chain are active in peptide synthesis. Single ribosomes are inactive.

Ribosomes are thus held together in groups or clusters in all organisms which have been studied—in reticulocytes of rabbits, in liver cells of rats, in the protoplasm of the slime mould, in human Hela cells, and in plant and bacterial cells. The electronmicroscope shows that, in these groups of ribosomes, or polysomes, the individual ribosomes are fastened together by a nucleic acid strand which is 10–15 Å in diameter. This strand is RNA; brief treatment with RNA-ase serves to cut it, releasing all the ribosomes as individual particles of sedimentation coefficient, 80 S, characteristic of single ribosomes of higher organisms. The average number of ribosomes per polysome, the number bound together by one RNA chain, varies from approximately 5 in Hela cells or approximately 4 in rabbit reticulocytes cells to 30 or 40 in virus-infected cells. There is however always a wide range in size. Even in, say, Hela cells, although the average number of ribosomes is 5 per polysome, there are present in the cells substantial numbers of polysomes with 2 to 30 ribosomes per group as may be visualized by electronmicroscopy, determined by sucrose density centrifugation and/or by analytical centrifugation.

But these are mere background facts. What can we say of ribosomal attachment to the messenger RNA chain? Why do we find more than one ribosome per messenger RNA chain? These matters have been studied most extensively by Goodman and Rich (1963) with the Hela cell ribosomal system, by Wettstein *et al.* (1963) with liver ribosomes, and particularly

TABLE 1

Attachment of P³²-labelled 80 S ribosomes to polysomes in vitro. A ribosomal preparation containing 12000 cpm is incubated for 5 min with a polysome preparation. Both are from reticulocytes

Conditions of incubation	Ribosomes, previously 80 S, bound to poly-somes after 5 min (cpm)
Complete system, inc. at 37°	1610
Complete system, inc. at 0°	245
No ATP, inc. at 37°	330

After Hardesty *et al.*, 1963.

elegantly by Schweet's group (1963) using the ribosomes of reticulocytes. Let us consider the experiments with reticulocytes.

Single ribosomes and polysomes are separated by sucrose density gradient centrifugation. Let us now add radioactively-labelled 80 S ribosomes to a reaction mixture containing unlabelled polysomes. The logic of the experiment is to use the great mass of the polysome as a sort of label for the messenger RNA and to determine whether or not 80 S ribosomes combine with such polysome-labelled messenger RNA. That ribosomes do attach to polysomes is shown in Table 1. In 5 minutes, about half of all of the polysomes in the reaction mixture have bound an 80 S ribosome. This attachment requires energy in the form of ATP and is enzymatic since it takes place minimally in the cold. That attachment occurs at the end of the messenger RNA chain is indicated by the finding of Goodman and Rich (1963) that binding of 80 S ribosomes to polysomes is proportional to the number of messenger RNA ends rather than to the total length of messenger RNA present. Finally, binding of ribosomes to polysomes is inhibited by the presence of small molecular weight poly A. The poly A acts, presumably, by being itself bound to the ribosome—thus blocking the messenger RNA attachment site. The poly A chains are sufficiently short so that two ribosomes cannot attach to one chain.

So much for attachment. The events following attachment of ribosome to polysome may be studied by similar methods. A

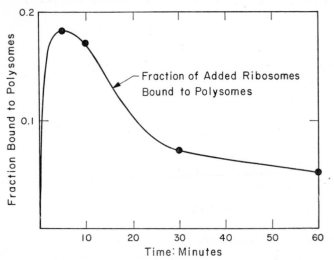

FIG. 6.1. Kinetics of binding of added 80 S ribosomes to pre-existing polysomes. The added 80 S ribosomes were P^{32}-labelled to distinguish them from the (unlabelled) ribosomes of the polysomes (after Hardesty *et al.*, 1963).

kinetic picture of the interaction of ribosome and polysome is shown in Fig. 1. The number of 80 S ribosomes newly bound to polysomes rises sharply to a maximum after 5 minutes at 37°. Thereafter it decays as the polysomes are degraded, reaching a low value after 60 minutes. It is a helpful feature of reticulocyte ribosomes that, *in vitro*, they can bind but once to messenger RNA. Once released, they cannot perform a second cycle of attachment and release. Bishop, Leahy, and Schweet have previously used this property of reticulocyte ribosomes in their classic work which demonstrates, by time sequential labelling, that the polypeptide chain grows from the N terminal end (Bishop *et al.*, 1960).

Thus 80 S ribosomes are sequentially bound to messenger RNA and again released. Further experiments demonstrate that attachment of 80 S ribosomes to polysome is associated with initiation of peptide chain growth from the N terminal end while release of the ribosome from the polysome is associated with release from the ribosome of a completed polypeptide chain. For this demonstration, two tools are used, the knowledge that binding of ribosome to messenger RNA is suppressed by poly A and the knowledge that the N terminal amino acid of haemoglobin, which is almost the only protein produced by reticulocytes is valine. Incorporation of valine into the N terminal position can therefore be used as the measure of initiation of peptide chain growth. Incorporation of amino acid into non-terminal positions can be used as a measure of chain growth itself. Finally, the appearance of labelled soluble haemoglobin can be used as the measure of peptide chain completion and release.

These experiments have been done in several ways. One way is to isolate, by sucrose density gradient centrifugation, those ribosomes which are associated with polysomes. The growing peptide chains of these ribosomes have been previously labelled *in vivo*—that is, before isolation of the polysomes from the reticulocyte cell. The polysomes are incubated with the ingredients necessary for protein synthesis and in addition, with poly A to prevent re-attachment of liberated ribosomes to native messenger RNA. The results of such an experiment (Fig. 2) show clearly that, with time, the labelled amino acid bound to ribosomes (in turn bound in polysomes) in growing peptide chains falls while the amount of labelled haemoglobin rises. The appearance of haemoglobin molecules completed during incubation is complementary to, and nicely balances, the disappearance of peptide chains attached to ribosomes. Parallel with the appearance of haemoglobin, is the appearance of free 80 S ribosomes which have been detached from the polysomal structure. In this type of experiment no opportunity is given for the initiation of new peptide chains. The complementary experiment which determines whether and how new peptide chains are initiated *in vitro*, must be done with C^{14}-labelled valine. Suppose that we incubate polysomes with C^{14}-labelled valine. Initiation of new peptide chains as measured by incorporation into the N terminal position does not occur.

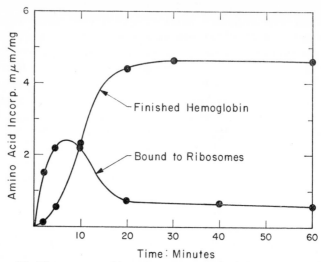

Fig. 6.2. Time course of incorporation of C14-labelled leucine into ribo-somally-bound ribosomes as well as the time course of appearance of free, finished haemoglobin. The incubation was carried out in the presence of poly A to suppress the initiation of new peptide chains (after Hardesty et al., 1964).

It occurs only if 80 S ribosomes are present in the system under conditions in which their attachment to polymers can also occur. In the experiment of Table 2, polysomes and 80 S ribosomes are used together with a complete protein-synthesizing reaction mixture including C14-labelled valine. We mix 80 S ribosomes and polysomes. Attachment occurs. Protein synthesis proceeds.

TABLE 2

Association of the initiation of new peptide chains (as measured by incorporation of N-terminal valine) with binding of ribosomes (previously 80 S) to polysome. Reticulocyte system in vitro. Inc. 60 min at 37°

System	Valine incorporated into peptide chains	
	Total ($\mu\mu$m)	% N-terminal
Complete	8·3	3·4
Complete + poly A	5·7	1·0

After Hardesty et al., 1963.

The data of Table 2 show that C^{14} valine is incorporated into protein and that a portion of it is N terminal. The rest of the valine is, of course, incorporated into growing peptide chains in non-terminal position. Uniformly-labelled haemoglobin contains 8·3 per cent of its valine in the N terminal position. In the experiment of Table 2, 3·4 per cent of the valine incorporated is in N terminal position. It may be concluded, therefore, that some 41 per cent of the labelled haemoglobin molecules were synthesized from start to finish during the *in vitro* experiment.

Table 2 shows also the influence of poly A upon the initiation of new peptide chains. The short-chain poly A, as we have seen, inhibits attachment of ribosome to polysome. The data of Table 2 show that, in the presence of poly A, the incorporation of valine into the N terminal position of haemoglobin peptide chains is suppressed. Such incorporation as does take place is almost entirely in other positions and is due to the completion of peptide chains previously initiated.

One final aspect of polysomal matters. If ribosomes are, in fact, bound together by messenger RNA, we would expect that when a cell is depleted of messenger RNA, the polysomes should gradually disappear and single ribosomes should appear in their place. This experiment has been done by, among others, Schlesinger (1963) with *Bacillus megaterium*. Pertinent data from Schlesinger's experiment are included in Table 3. The logic of the experiment consists in determination of the partition

TABLE 3

Decay of polysomes present in cells of Bacillus megaterium as a result of decay of messenger RNA. Production of fresh messenger RNA *blocked by actinomycin-*D

System	% of ribosomes in form of		Rate of protein synthesis (cpm/1 min)
	polysomes	single ribosomes	
Complete	49	51	16600
Complete but 3 min after administration of actinomycin-D.	18	82	6050

After Schlesinger, 1963.

of ribosomes between polysomes and single ribosomes both in normally growing cells and in cells which have been treated with actinomycin D (which specifically blocks DNA-dependent RNA synthesis) for a time sufficient to substantially lower their concentration of messenger RNA. For *B. megaterium,* such a time is 3 minutes. The data of Table 3 show that in normal *B. megaterium* cells, about one half of all ribosomes are bound in polysomes—the other half being present as free single particles. After a 3-minute treatment with actinomycin D, on the contrary, over 80 per cent of the ribosomes are present as free single particles and only a small minority are still present as polysomes. The data of Table 3 also show that the ability of the cell to support protein synthesis is closely correlated with the number of ribosomes still bound in polysomes. The suppression of RNA synthesis which results in gradual depletion of messenger RNA not only results in transfer of ribosomes from polysome to ribosome category but in addition causes rate of protein synthesis to drop correspondingly.

What is lacking in our picture of the interaction of ribosome and messenger RNA is direct evidence that the ribosome progresses serially down the messenger RNA molecule as it decodes the message contained therein, as well as direct evidence that the end of the messenger RNA from which the ribosome is released is the end opposite that from which it started its journey. Nonetheless the circumstantial evidence is impressive. We have seen that ribosomes combine with messenger RNA at chain ends only, while after combination of ribosome with messenger RNA, we find ribosomes at various positions along the length of the messenger molecule. As synthesis of the peptide chain is finished ribosomes leave the polysome. We possess then, every link in the chain of evidence required to establish that ribosomes do, in fact, read the messenger RNA sequentially and that attachment and release of ribosome to messenger are associated respectively with initiation and completion of peptide chain growth.

SELECTED REFERENCES

Concept of polysomes

BARONDES, S. and NIRENBERG, M., *Science* **138,** 813 (1962).

SPYRIDES, G. and LIPMAN, F., *Proc. Nat. Acad. Sci., Wash.* **48,** 1977 (1962).

WARNER, J., RICH, A. and HALL, C., *Science* **138,** 1399 (1962).

SLAYTER, H., WARNER, J., RICH, A. and HALL, C., *J. mol. Biol.* **7,** 652 (1963).

Polysomes in relation to protein synthesis

WARNER, J., KNOPF, P. and RICH, A., *Proc. Nat. Acad. Sci., Wash.* **49,** 122 (1963).

GIERER, A., *J. mol. Biol.* **6,** 148 (1963).

WETTSTEIN, F., STAEHLIN, T. and NOLL, H., *Nature, Lond.* **197,** 430 (1963).

GOODMAN, H. and RICH, A., *Nature, Lond.* **199,** 318 (1963).

GILBERT, W., *J. mol. Biol.* **6,** 374 (1963).

Detailed analysis of ribosome-polysome kinetics, etc.

HARDESTY, B., ARLINGHOUSE, R., SCHAEFFER, J. and SCHWEET, R., in *Synthesis and Structure of Macromolecules.* Cold Spring Harbor Symp. in Bio., Vol. 28 (1963).

HARDESTY, B., MILLER, R. and SCHWEET, R., *Proc. Nat. Acad. Sci., Wash.* **50,** 924 (1963).

HARDESTY, B., HUTTON, J., ARLINGHOUSE, R. and SCHWEET, R., *Proc. Nat. Acad. Sci., Wash.* **50,** 1078 (1963).

SCHLESINGER, D., *J. mol. Biol.* **7,** 569 (1963).

Growth of peptide chain from N terminal end

BISHOP, J., LEAHY, J. and SCHWEET, R., *Proc. Nat. Acad. Sci., Wash.* **46,** 1030 (1960).

DINTZIS, H. M., *Proc. Nat. Acad. Sci., Wash.* **47,** 247 (1961).

FURTHER FUNCTIONS OF THE NUCLEUS

WE HAVE seen that the chromatin of the nucleus conducts DNA-dependent RNA synthesis and that the RNA thus generated includes the messenger RNA's transcribed from those genes which are not repressed. Our attention has in fact been wholly focused upon the production of messenger RNA and we have deliberately ignored the fact that the nucleus possesses other functions. These include the production of further classes of RNA, classes which differ from messenger RNA in physical nature, in base composition, in kinetics of biogenesis and of destruction, and in cellular function. We will ultimately talk about a possible new class of nuclear RNA. Before we can properly discuss a new class of RNA, we must know what there is to know about the classical ones. Let us therefore consider the nucleus as a whole. There is a very large literature on the nucleus. This discussion will be principally confined to our own work on the nuclei of plant cells.

To study nuclei, we must first isolate them from the cell. To do this we must first rupture the cell membranes so that the nuclei can fall out. The rupturing of cell membranes is commonly accomplished by some species of grinding machine such as a homogenizer or blendor. The blendor ruptures by shearing cells. The nucleus is a large and delicate structure and the shear gradients required to rupture cell membranes are, in general, and for plant cells in particular, sufficient to rupture nuclear membranes as well. This is the fact upon which the procedures for direct isolation of chromatin are based. For the isolation of intact nuclei, we must therefore rupture cells more gently. To this end we have developed a machine—the low shear tissue disruptor shown in Fig. 1 (Rho and Chipchase, 1962). It consists first of a stationary bed along which moves a belt of nylon mesh. The tissue of interest, for example, young pea seedlings, is dropped upon the moving belt and is carried between two counter-rotating rollers. The tissue is compressed. Let us imagine the fate of an individual cell as the tissue passes between the rollers. As it enters the rollers, one end of the cell becomes compressed. Since the liquid contents of cells are essentially incompressible, the far end of the cell, not yet between the rollers, expands because it is subjected to a hydro-static pressure. The force with which the rollers press upon the

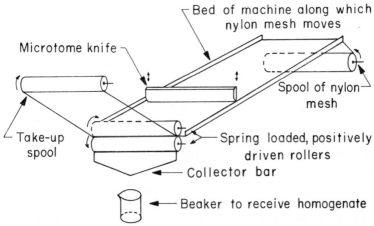

FIG. 7.1. The enuclear reactor for low shear tissue disruption. The tissue is placed upon a nylon mesh belt which moves from right to left down the bed of the machine. A cutter bar (microtome knife) chops the tissue into 1-mm long fragments. The tissue than passes through the rollers. Cell wall fragments remain on the nylon and accompany it to the takeup roller. The homogenate is collected in a beaker below the rollers (after Rho and Chipchase, 1962).

cell between them is adjustable by spring loaded set screws. It has been cleverly set so that the pressure upon the far end of the cell is just sufficient to burst open the cell membrane without disrupting the nuclear membrane. The contents of the cell filter through the meshes in the nylon belt and through an appropriate collecting device into a refrigerated container. The cell membranes remain on the nylon belt, never to bother the investigator again. Since nuclei expressed from cells into cell juice ordinarily aggregate, a medium to prevent this is dispensed over the tissue just as it enters the rollers. This medium contains sucrose (0·25 M), tris buffer (0·006 M, pH 7·2), and calcium chloride (0·003 M). In this medium nuclei remain stable and non-aggregated over periods of several hours.

The cell contents, as they are liberated from the rupturing of individual cells, must travel through the tissue until a cut end is reached. A rapidly chopping microtome blade is therefore mounted above the nylon belt and chops the tissue into one-millimeter segments—a length optimal for recovery of intact nuclei from pea stems. In this machine then, tissue is disrupted under exceedingly low shear. Nuclei are liberated intact and in good yield. The device is of high capacity—the model shown in Fig. 1 can deal with 1 kilogram fresh weight of tissue in 15 minutes. To this fine device, we have given a name—any device so novel and effective deserves a name. We call it the 'enuclear reactor'.

The further purification of nuclei is simple. They are centrifuged from the tissue homogenate at 350 × g. They are further purified by centrifugation through a 0·25 to 1·2 M sucrose density gradient—centrifugation being at 450 × g for 10 minutes (Rho and Chipchase, 1962).

The study of nuclear affairs requires also that we be able to separate the subnuclear components. This matter has been studied in our group by Rho, Birnstiel, Chipchase, and Hyde among others. Having, with such patience and craft, isolated intact nuclei, we proceed at once to disrupt them. This is accomplished by first removing the calcium ions from the nuclear suspension by citrate, since in the presence of calcium ions, the nuclear membranes are tough and not readily disruptable. The nuclei are next blended for 1 minute at 40 volts in sucrose of density 1·3. By such blending in viscous medium, the nuclear membranes are largely disrupted but the nucleoli, the largest and most obvious nuclear substructures, are left intact. The density of the nuclear homogenate is now reduced to 1·22 and the nucleoli removed by centrifugation for 20 minutes at 20000 × g. The density of the nucleolar supernatant is further lowered to 1·04 and the chromatin removed by centrifugation at 20000 × g for 15 minutes. From the chromosomal supernatant, ribosomes are next pelletted by centrifugation for 2 hours at 105000 × g. Soluble nuclear enzymes remain in the supernatant. Separation of the nuclear subsystems is merely separation of the cellular subsystems revisited but on a smaller scale.

The principal nuclear subcomponents are then available for chemical and biological study. Let us consider the functions of the several nuclear subsystems.

Chromatin, of course, supports the synthesis of RNA. A portion of this as we have seen is messenger RNA for the synthesis of specific proteins. A further portion is the transfer RNA which transports amino acids from activating enzymes to ribosomes and messenger RNA. The synthesis of transfer RNA by chromatin, is inhibited by actinomycin D and is therefore DNA-dependent (Chipchase and Birnstiel, 1963). There are therefore genes responsible for the genesis of transfer RNA as there are also in micro-organisms (Goodman and Rich, 1962). The work on the nuclear genesis of transfer RNA is elegant and complete. Let us, however, omit here the details of the study of transfer RNA in favour of the nucleolus. For what purpose is the nucleolus in the nucleus?

It has been known since early times (Sirlin, 1960) that when cells are incubated with labelled precursors of RNA the nucleolus as well as the chromatin of the nucleus become labelled as shown by radioautography. It was in fact once believed that the nucleolus was the principle seat of nuclear RNA synthesis. This is, however, clearly not the case. Not only do isolated nucleoli have little ability for RNA synthesis, but in addition, the RNA of the nucleolus becomes labelled only after that of the extranucleolar nucleoplasm. This has been found both by radioautography and by nuclear fractionation studies. Thus Woods (1959) has shown for *Vicia* roots by radioautography that the RNA of the extranucleolar nucleoplasm attains steady state labelling in 14 minutes but that 1·5 hours are required for the nucleolus to attain its final steady state of labelling. Similar conclusions have been reached by Goldstein and Micou (1959). Another way of approaching the matter (Rho and Bonner, 1961) has been to allow isolated nuclei to become labelled for various lengths of time, using a suitable labelled precursor of RNA, and to then determine separately the amounts of labelled RNA in the several subnuclear components. The results are clear-cut. Over short time periods, 1 to 4 minutes, nearly all of the newly synthesized RNA is found in the chromatin of the nucleoplasm and little is found in the nucleolus. Over longer times, the nucleolar RNA also becomes labelled. Pulse-chase experiments in which a labelled precursor of RNA is supplied to isolated nuclei for a brief time (15 minutes) and then diluted by a thousandfold excess of the unlabelled precursor make it possible to follow the fate of the labelled chromosomal RNA. A portion of this

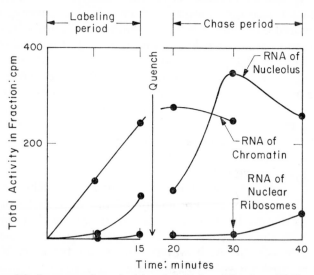

FIG. 7.2. Incorporation of tritium-labelled cytidine into RNA of nuclear subcomponents of pea-embryo nuclei. In this pulse-chase experiment, isolated nuclei are first incubated in the labelled precursor for 15 minutes. Incorporation into RNA is then quenched with a 1000-fold excess of unlabelled CTP. In the succeeding chase period the fate of the previously labelled RNA is followed (after Rho and Bonner, 1961).

material appears transitorily in the nucleolus. With the further passage of time, the RNA is again lost to the nucleolus only to reappear after 30 to 40 minutes in the ribosomes of the nucleoplasm (Fig. 2).

That ribosomes occur in the nucleus was first shown by Ts'o and Sato in 1959, who isolated them from nuclei and showed, by physical and chemical characterization, that they are identical with cytoplasmic ribosomes. Ts'o and Sato further found in short term labelling experiments with P^{32}-labelled phosphate that not only do the nuclear ribosomes become labelled, but that the specific activity of their RNA is much greater than that of the cytoplasmic ribosomes. It would appear then, that the nuclear ribosomes are those ribosomes whose RNA has been most recently synthesized. The fractionation of subnuclear components at varying times during the course of pulse-chase experiments has further shown that the RNA which appears in cytoplasmic ribosomes passes through the nucleolus in its journey from chromatin where it is generated to its ribosomal destination (Rho and Bonner, 1961).

Ribosomes then clearly have a nuclear origin and in this origin the nucleolus appears to play a role. As a prelude to discussion of the role of the nucleolus in ribosomal matters, let us first discuss the properties of the ribosome itself. Pure ribosomes were prepared by Ts'o and his colleagues in 1955 and 1956 and by Chao and Schachman in the same years. The ribosomes of higher creatures, plants and animals alike, have a sedimentation coefficient of 80 S and thus differ from those of bacteria, which have been found to possess a sedimentation coefficient of 70 S. The sedimentation coefficient of 80 S corresponds to a molecular or particle weight of 4×10^6. Of this mass, very nearly one half is RNA and the remainder protein. Furthermore, ribosomes are made of subunits and these are stapled together by magnesium ions. Removal of magnesium causes ribosomes to dissociate reversibly into two fragments—one of two thirds the original weight, the so called 60 S subunit and the other of one third the original weight—the so called 40 S subunit. (The corresponding subunits of bacterial ribosomes have sedimentation coefficients of 50 S and 30 S.) These qualities then are the stigmata of the ribosome; a sedimentation coefficient of 80 S, a composition of one-half RNA one-half protein and dissociation into characteristic subunits upon the removal of magnesium ions (Fig. 3). Ribosomes have other properties too; for example, once made, they live for a very long time. Perhaps most characteristic of all are the sedimentation coefficients of the ribosomal RNA. The 60 S subunit, yields a pure RNA of sedimentation coefficient 28 S while the 40 S subunit yields a pure RNA of sedimentation coefficient 18 S (Ts'o et al., 1956, 1958). Since ribosomal RNA is long lived and since there are many ribosomes in a cell, it is not surprising that the great bulk of cellular RNA consists of 28 S and 18 S components.

The protein component of ribosomes too has its own unique characteristics. Thus the amino acid composition of ribosomal protein is characterized by high contents of aspartic and glutamic acids as well as by high contents of lysine and arginine and this is true of the protein of all ribosomes thus far investigated.

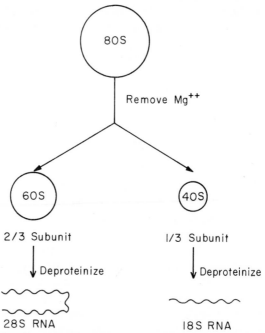

Fig. 7.3. Dissociation of 80 S Ribosome to 60 S and 40 S subunits by removal of Mg ions and characterization of the RNA contained in the two types of subunits. The 28 S and 18 S RNA molecules are also obtained by direct deproteinization of the 80 S ribosome.

Ribosomal protein is, however, clearly distinguishable from histone by its lower content of basic amino acids, by the presence of tryptophane, and by its relative insolubility in acid.

Let us now return to the nucleolus. Although this structure is relatively inert in making RNA it is exceedingly active in protein synthesis (Birnstiel *et al.*, 1961). Indeed in short time incubations of isolated nuclei, protein synthesis is essentially confined to the nucleolus, a fact indicated also by radioauto-graphic procedures (Sirlin, 1960). In pulse-chase experiments, the protein originally synthesized in the nucleolus disappears with time only to reappear in the nucleoplasm. What species of protein is this? This question has been answered directly by Birnstiel and Hyde (1963). They allow isolated nuclei to synthesize protein in a medium including the four riboside triphosphates and the twenty amino acids—one labelled. After a few minutes of incubation, the nuclei are fractionated into the subnuclear components. From the isolated nucleoli, separate classes of protein are removed and the amino acid composition of each determined. One class of nucleolar protein is the acid soluble histone, which as we will see in Chapter 8 is nucleolar in origin. But the great bulk of the newly synthesized nucleolar protein is contained in an acid insoluble fraction with the amino acid composition of ribosomal protein. Can it be then that the ribosomes are manufactured in the nucleolus?

That ribosomes are contained in the nucleolus was first indicated by electron microscopy. Birnstiel and Hyde (1963) find for example that the nucleolus contains a mass of particles of diameter approximately 200 Å which stain with lead as do ribosomes. This is certainly a straw in the wind. But more rigorous is the isolation and physical characterization of nucleolar RNA by Chipchase and Birnstiel (1963). The deproteinized RNA of isolated nucleoli largely consists, as is shown in Fig. 4, of the 28 S and 18 S components characteristic of ribosomal RNA. Birnstiel, Chipchase, and Hyde (1963) have also found that ribosomes can themselves be prepared from isolated nucleoli. In the nucleolus, ribosomes are bound, apparently, to membranes, just as in the cytoplasm they are bound to the endoplasmic reticulum. As in the latter case, ribosomes may be liberated from the nucleolus by judicious extraction with a weak detergent—deoxycholate. The sedimentation profile of such solubilized nuclear material reveals the presence of not only 80 S ribosomes but also of 60 S and 40 S subunits. The 80 S particles may be reversibly dissociated, by removal of magnesium, to 60 S and 40 S subunits, just as with cytoplasmic ribosomes.

Very well, the nucleolus receives certain species of RNA from the chromatin and clothes it in ribosomal protein. The RNA thus clothed possesses the physical properties of ribosomal RNA. But there is a still more rigorous way to demonstrate that the nucleolar RNA is identical with that of the cytoplasmic ribosomes. The method of choice is the DNA-RNA hybridization technique of Hall and Spiegelman (1961). By this technique, we can determine whether a given RNA is complementary to any given DNA and hence could have been formed from it by transcription. It is also possible to find out whether

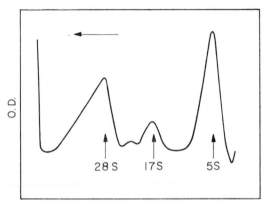

FIG. 7.4. Sedimentation analysis of nucleolar RNA. The extracted RNA was overlayered on a solution of 2 M CsCl and centrifuged at 56000 rpm. The horizontal arrow represents the direction of the centrifugal field. Pattern made by standard ultraviolet optics.

The 28 S and 18 S components constitute 50–60 per cent of the nucleolar RNA. The 5 S peak is due to transfer RNA. A minor 23 S component is evident and was found in all experiments (after Birnstiel *et al.*, 1963).

two samples of RNA are similar to, or different from one another in base sequence and hence in nature of the information encoded in them.

The logic of the hybridization technique is simple. First we transform double-stranded DNA to single strands by heating it above its T_M and then quickly cooling. Quick cooling of the melted DNA ensures that the two complementary strands have no opportunity to again seek and double strand with one another. To the single-stranded DNA we add our RNA. We then incubate the mixture at a sufficiently high temperature to permit annealing of RNA to complementary DNA but insufficient to permit annealing of complementary DNA chains to one another. During annealing RNA chains seek and mate in double helical conformation with DNA chains complementary to themselves. After a sufficiently long annealing period (2 hours at 63°) any unhybridized RNA is removed by treatment with RNA-ase to which the DNA-RNA hybrid is totally resistant. The oligonucleotides released are removed by filtration on an appropriate millipore filter. We then determine the amount

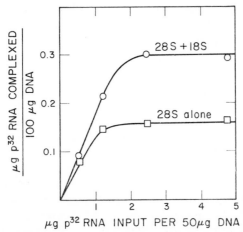

FIG. 7.5. Saturation curves for hybrid formation between denatured pea DNA and ribosomal RNA subunits. In each case 50 μg of DNA was annealed with varying amounts of P³²-labelled RNA and the hybrids assayed as described in Methods (after Chipchase and Birnstiel, 1963).

of RNA bound in hybrid form. This is usually done by starting with isotopically labelled RNA so that the hybridized RNA may be determined by its radioactivity.

Application of the procedure requires that we first make a saturation curve to determine how much input RNA is required to saturate all complementary sites in the DNA sample. From the amount of RNA bound at saturation by a known amount of DNA, we find what proportion of the DNA is involved in the making of this particular kind of RNA (Fig. 5).

Suppose that we now wish to discover whether two samples of RNA are identical or different in base sequence. Two procedures are available. In the first, both samples of RNA must be labelled. We compare the amount of RNA which is bound at saturation of the DNA when only one kind of RNA is added with the amount bound to the same weight of DNA when both are added. If the two samples of RNA have identical base sequence, then the amount bound in hybrid form will be no greater when both kinds are added than it is when only one kind is used. Adding the second kind of RNA is no different from adding more of the first kind, of which there is already a saturating amount. If however, the two samples of RNA differ in base sequence—were transcribed originally from different portions of the DNA and are the products of different genes—than, the amount of RNA hybridized by a given amount of DNA when both kinds are present will equal the sum of the amounts hybridized separately. The different kinds of RNA molecules will seek and hybridize with different portions of the DNA of the genome.

In the second procedure, one RNA sample must be labelled, the other not. We hybridize one DNA sample with a saturating amount of the labelled variety of RNA and another with this same saturating amount of labelled RNA but in the presence of an equal amount of the unlabelled RNA. If the two kinds of RNA are identical in base sequence then they will seek to hybridize with the same DNA stretches. They will compete with one another on a equal basis. The amount of labelled RNA bound in hybrid form will be halved by the presence of an equal amount of unlabelled identical RNA. If, on the other hand, the two RNA samples differ in base sequence they will not compete. The amount of the labelled variety bound in hybrid form by a standard amount of DNA will not be influenced by the presence of the unlabelled RNA. The two kinds of RNA again will seek and hybridize with different stretches of the DNA; with different genes.

These hybridization procedures have been used by Chipchase and Birnstiel (1963) for study of the identity of the nucleolar RNA. They determined whether nucleolar RNA is identical with or different from the RNA of the cytoplasmic ribosomes. They used the second of the procedures described above. The 28 S and 18 S RNA components were prepared, unlabelled, from nucleoli. Labelled 28 S and 18 S RNA components were prepared from cytoplasmic ribosomes. To this end tissue was incubated for 24 hours in P^{32}-labelled orthophosphate. The P^{32} was then washed out with a 10000-fold excess of unlabelled orthophosphate and incubation continued for a further 12 hours. During this period, any unstable messenger RNA, was given an opportunity to be degraded. Ribosomes were then isolated and repeatedly washed by centrifugation at 105000 × gravity. Ribosomal RNA, extracted with phenol was then separated into 28 S and 18 S subunits by centrifugation in a sucrose density gradient. The individual RNA's were further freed of any DNA by pelleting in CsCl of density 1·73 in which DNA floats while RNA sinks. Whole genomal DNA was prepared from whole nuclei while nucleolar DNA—about one-seventh of the total genome—was prepared from isolated nucleoli.

Saturation curves for cytoplasmic ribosomal RNA with whole genomal DNA are shown in Fig. 5. It is clear that at saturation very close to 0·3 per cent of the total genomal DNA is complexed with cytoplasmic ribosomal RNA. Such hybrid formation is quite specific, as is shown in Table 1. No hybrid

TABLE 1

Hybridization of pea DNA with P³²-labelled pea cytoplasmic ribosomal RNA in the presence or absence of other heterologous but nonlabelled RNA's

Source of DNA	Source of RNA added to system (P³²-pea ribosomal RNA : un-labelled heterologous RNA)	μg P³² ribo-somal RNA hybridized† per 100 μg DNA
Pea	P³²-pea ribosomal (28 S 18 + S) only	0·29
Phage T4	P³²-pea ribosomal (28 S 18 + S) only	0·02
Pea	P³²-pea ribosomal + unlabelled coli ribosomal RNA (4X)	0·31
Pea	P³²-pea ribosomal + unlabelled pea ribosomal (1X)	0·16
Pea	P³²-pea ribosomal + unlabelled pea ribosomal (4X)	0·08
Pea	P³²-ribosomal + unlabelled pea nucleolar RNA (4X)	0·10

† An amount of RNA sufficient to cause saturation of the DNA was used in all cases.
After Chipchase and Birnstiel, 1963.

6

is formed, for example, between cytoplasmic ribosomal RNA and DNA of bacteriophage T_4, a creature that does not know how to make ribosomes. In this vein too, hybridization of pea cytoplasmic ribosomal RNA with pea DNA is not interfered with in the slightest by the presence of *E. coli* ribosomal RNA. The RNA's of the ribosomes of these two creatures therefore differ from one another. The presence of nucleolar RNA does, however, inhibit complex formation between cytoplasmic ribosomal RNA and whole genomal DNA. Equal amounts of

TABLE 2

Hybridization of pea cytoplasmic ribosomal RNA with whole pea nuclear DNA and with pea nucleolar DNA

Source of DNA	Source of RNA	μg RNA hybridized? per 100 μg DNA
Whole pea nucleus	ribosomal (28 S + 18 S)	0·30
Whole pea nucleus	ribosomal (28 S)	0·17
Pea nucleolus	ribosomal (28 S + 18 S)	0·32
Pea nucleolus	ribosomal (28 S)	0·19

† An amount of RNA sufficient to cause saturation of the DNA was used in all cases.
After Chipchase and Birnstiel, 1963.

these two RNA's exactly halve hybrid formation by cytoplasmic ribosomal RNA. We must conclude that nucleolar RNA and cytoplasmic ribosomal RNA are not only similar in physical properties, but are also identical in base sequences.

We have already seen that nucleolar preparations always contain DNA. Might this DNA contain the genes responsible for the genesis of ribosomal RNA? To test this point DNA was prepared from nucleoli and its ability to hybridize with ribosomal RNA determined. The data of Table 2 show that, although nucleolar DNA does hybridize with ribosomal RNA, it is no more effective in this function than whole genomal DNA. We must conclude that although some of the genes which make ribosomal RNA are contained in the DNA which accompanies the nucleolus, others are contained in the genome at large.

We conclude then, that the RNA of ribosomes is formed by DNA-dependent RNA synthesis on the chromosome. The RNA thus synthesized moves in some way, not known, to the nucleolus. There it is clothed in ribosomal protein. We find, in the nucleolus a heterogeneous mixture of 80 S ribosomal particles, 60 and 40 S subunits and ribosomal protein not clearly associated with such subunits; subunits perhaps in the throes of being born. These nucleolar ribosomes, as they are being fabricated in the nucleolus are themselves incapable of conducting protein synthesis as can the finished ribosomes of the cytoplasm or even of the nucleoplasm. (Birnstiel *et al.*, 1963). Nucleolar ribosomes appear to be truly ribosomes under assembly and not yet ribosomes the engines of protein synthesis.

What system in the nucleolus conducts the synthesis of ribosomal protein? Are there ribosomes in the nucleolus for the synthesis of ribosomal protein? What messenger RNA do the ribosomes which assemble the proteins of ribosomes use? Does the messenger function reside in the ribosomal RNA itself? These are all questions for the future—important questions. There is but one further fact to put forward—not in answer to, but in comment on these questions. The protein-synthesizing function of the nucleolus is due to something bound in its structure. It is a something not readily washed out or eluted during nucleolar isolation (Birnstiel and Hyde, 1963).

It may be cause for some surprise that the production of ribosomal RNA consumes as much as 0·3 per cent of the total genomal DNA. Ribosome-making is an important task, but we might have predicted it to be less important than this figure suggests. A smaller figure, about 0·03 % has in fact been found for human Hela cells (McConkey and Hopkins, 1964). In any case, there must then be a multiplicity of genes for the production of ribosomal RNA and we must anticipate that the RNA's of different ribosomes will prove to be different from one another because mutation will have caused the many and originally similar genes for ribosomal RNA making to become different from one another in the course of time. This would appear to be true also in *E. coli* for which Yanofsky and Spiegelman (1962) have shown that 0·3 per cent of the genome is concerned with the formation of ribosomal RNA.

We see then that the nucleolus is the engine of ribosomal synthesis and specifically of the synthesis of ribosomal protein and of ribosomal assemblage. We have gone at this matter the hard way however. Brown and Gurdon (1964) have found a much simpler and more elegant route to the same conclusion. They have made use of a mutation of *Xenopus*—the South African clawed toad which, when homozygous, abolishes nucleoli. True, this mutation in homozygous form is ultimately lethal but nonetheless, the embryo survives for a time because it inherits ribosomes from the ovule. Anucleolate *Xenopus* embryos cannot make ribosomes. Thus simply is the role of the nucleolus established by Gurdon and Brown.

SELECTED REFERENCES

Low shear tissue disruptor (the enuclear reactor)

RHO, J. H. and CHIPCHASE, M., *J. cell. Biol.* **14,** 183 (1962).

Isolation of nuclei from plant cells and separation of nuclear subsystems

BIRNSTIEL, M. CHIPCHASE, M. and BONNER, J., *Biochem. biophys. Res. Commun.* **6,** 161 (1961).
RHO, J. H. and BONNER, J., *Proc. Nat. Acad. Sci., Wash.* **47,** 1611 (1961).

Intranuclear seat of RNA *synthesis—radioautography*

SIRLIN, J., *Expl Cell Res.* **19,** 29 (1960).
WOODS, P., *Brookhaven Symp. Biol.* **12,** 153 (1959).
GOLDSTEIN, L. and MICOU, J., *J. Biochem. biophys. Cytol.* **6,** 1 (1959).

Intranuclear seat of RNA *synthesis—isolation of nuclear subsystems*

RHO, J. H. and BONNER, J., loc. cit.
CHIPCHASE, M. and BIRNSTIEL, M., *Proc. Nat. Acad. Sci., Wash.* **50,** 1101 (1963).
BIRNSTIEL, M., RHO, J. and CHIPCHASE, M., *Biochem. biophys. Acta* **55,** 734 (1962).

Origin of transfer RNA

CHIPCHASE, M. and BIRNSTIEL, M., *Proc. Nat. Acad. Sci., Wash.* **49,** 692 (1963).
GOODMAN, H. and RICH, A., *Proc. Nat. Acad. Sci., Wash.* **48,** 2101 (1962).

Intranuclear seat of protein synthesis

BIRNSTIEL, M., CHIPCHASE, M. and BONNER, J., loc. cit.

SIRLIN, J., in *The Cell Nucleus* (J. S. MITCHELL, Ed.) p. 35. Butterworths, London (1960).

BIRNSTIEL, M., CHIPCHASE, M. and HAYES, R., *Biochem. biophys. Acta* **55,** 728 (1962).

FLAMM, W. G., BIRNSTIEL, M. and FILNER, P., *Biochem. biophys. Acta* **76,** 110 (1963).

Nature and origin of ribosomes

TS'O, P. O. P. and SATO, C. J., *expl Cell Res.* **5,** 19 (1959).

BONNER, J., in *Protein Synthesis* (R. HARRIS, Ed.).   Academic Press London (1960).

BIRNSTIEL, M. CHIPCHASE, M. and HYDE, B., *Biochem. biophys. Acta* **76,** 454 (1963).

BIRNSTIEL, M. and HYDE, B., *J. cell Biol.* **18,** 41 (1963).

CHIPCHASE, M. and BIRNSTIEL, M., *Proc. Nat. Acad. Sci., Wash.* **50,** 1101 (1963).

TS'O, P. O. P. *Ann. Rev. Pl. Physiol.* **13,** 45 (1962).

Hybridization of RNA with DNA—technique

HALL, B. and SPIEGELMAN, S., *Proc. Nat. Acad. Sci., Wash.* **47,** 137 (1961).

Hybridization of RNA with DNA—application to ribosomal RNA

YANOFSKY, S. and SPIEGELMAN, S., *Proc. Nat. Acad. Sci., Wash.* **48,** 538 (1962). ibid **48,** 1069, 1962, ibid **48,** 1466, 1962

CHIPCHASE, M. and BIRNSTIEL, M., *Proc. Nat. Acad. Sci., Wash.* **50,** 1101 (1963).

MCCONKEY, E. H. and HOPKINS, J. W., *Proc. Nat. Acad. Sci., Wash.* **51,** 1197 (1964).

Nucleolusless mutant of Xenopus

BROWN, D. and GURDON, J., *Proc. Nat. Acad. Sci., Wash.* **51,** 139 (1964).

ELSDALE, T. R., FISCHBERG, M., and SMITH, S. *Expl. Cell. Res.* **14,** 642 (1958).

8

PROPERTIES OF THE HISTONES

WE HAVE seen in several contexts that the histone component of the chromosome is either itself the agent of, or is clearly associated with the agent of, genetic repression. We take chromatin in which a particular gene is repressed, remove the histone and that gene is derepressed. We must conclude that histones are interesting. Let us therefore consider the histones.

The discovery of the histones, as well as of DNA itself, is due to Miescher almost one hundred years ago. Miescher, who worked with pus cells which he soaked up from infected wounds, realized that histones are somehow closely associated with the DNA of the cell. Following Miescher, Kossel served as first chemist of the histones and of the related protamines, and Kossel's little book *The Histones and Protamines* has been for nearly forty years, the only book devoted to these interesting proteins. To Kossel is due our appreciation of the fact that the histones and protamines contain a large proportion of basic amino acids. The successor to Kossel as the great champion of the histones has been Stedman of the University of Edinburgh, who showed us first that the histones are a group of proteins rather than a single protein. Stedman devised procedures by which histones may be fractionated, and was also the first to propose a function for the histones within the nucleus. He proposed that the histones function in some way in controlling the expression of genetic activity. This proposal, made in 1951, could not at that time be supported with experimental evidence; the methodology for doing so was not yet available. It may be remarked in passing, however, that Stedman looks with favour upon the present developments concerning the histones. Histones are basic proteins which are associated with DNA. They are thus to be distinguished from the superficially similar basic proteins which are associated with RNA in the structure of the ribosome. If we wish to prepare histones and wish to assure ourselves that we obtain only histones, the best way is to first prepare chromatin by the methods described in previous chapters. Protein may then be dissociated from DNA either by use of high ionic strength, or by 0·1–0·2 M acid, an acidity sufficient to repress the dissociation of the phosphate groups of DNA but low enough so that the histones remain soluble. Most workers have not however, started with chromatin as their raw material, but have rather extracted whole nuclei or even whole tissue rich in nuclei with salt or acid. Among all starting materials the most fashionable has been the calf thymus. Calves not only have large thymus glands, but, in addition, these glands consist of cells in which the bulk of the volume is nucleus. Most histone chemists therefore merely extract whole thymus glands with acid. Nonetheless, the histones thus obtained closely resemble those obtained by more rigorous procedures. We will first discuss the histones of calf thymus upon which so much work has been done and then proceed to consider how far they resemble those of other cells, tissues, organs, and creatures.

The first result of histone chemistry is that the whole histone of calf thymus chromatin is a mixture. This is shown for example by the data of Fig. 1. For this experiment the whole acid soluble extract of calf thymus chromatin has been separated by column chromatography on Amberlite IRC 50. Proteins have been eluted from the column with a gradient of increasing concentration of guanidinium chloride according to the procedures of Rasmussen et al. (1962). The data of Fig. 1 show that the whole calf thymus histone is separated into four major fractions to which are assigned numbers in order of their appearance. Histone fraction I is in turn incompletely resolved to two subcomponents, histone fractions Ia and Ib. Next appears histone fraction II, followed by histone fractions III and IV. The data of Fig. 1 show also how fractions I and II constitute the great bulk of the histone of calf thymus, 80 per cent or more of the total. Similar separations of calf thymus histone into fractions I to IV can be achieved by other methods and in fact a whole range of procedures have been developed. Amongst the methods which are most in use today is that of Johns and Butler (1962) in which whole thymus histone or even whole thymus tissue is extracted with 5 % perchloric acid. Histones I and II are soluble, III and IV are not. The mixture of fractions I and II is precipitated from the perchloric acid extract with ethanol. The perchloric acid insoluble histones III and IV are then dissolved in 0·2 N sulphuric acid and separated from one another with a pH gradient on a carboxymethyl cellulose column. This column may also be used to separate all

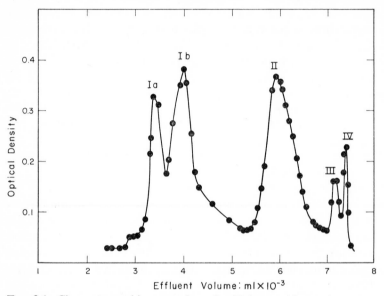

FIG. 8.1. Chromatographic separation of calf thymus histone into component fractions. Protein concentration followed by optical density at 400 mμ of turbid solution resulting from treatment of 0·5-ml of sample with 2·5 ml of 1·1 M trichloracetic acid. Amberlite IRC 50 column and gradient of elution with guanidinium chloride, 8 to 40 per cent (after Murray, 1964).

four classes of histones by eluting with a pH gradient from pH 4·5 decreasing gradually to pH 1. But the point of importance in the present context is that a variety of methods lead to the separation from calf thymus histone of four principal fractions. Let us now consider what it is which makes these histone fractions different from one another.

The amino acid compositions of the principal histone fractions of calf thymus are presented in Table 1. The first and most striking point is that they contain neither cysteine, cystine, nor tryptophane. They are all poor also in the aromatic amino acids, tyrosine and phenylalanine. From the data of Table 1 also emerges the most spectacular and characteristic difference in amino acid composition between the histone fractions. Although the total lysine plus arginine content of all histones is approximately 24–28 mole per cent, the relative proportions

TABLE 1

Amino acid compositions of histones of calf thymus as separated by chromatography with guanidinum chloride

Amino acid	Histone fraction				Whole thymus histone
	Ib	IIb	III	IV	
Lys	26·2	13·5	9·4	9·0	16·0
His	0·2	2·8	1·6	1·6	1·5
Arg	2·2	7·9	12·9	12·8	7·9
Asp	2·5	5·6	4·4	4·5	4·7
Thr	5·4	5·2	7·4	7·4	5·5
Ser	6·5	7·0	4·1	4·1	5·2
Glu	4·3	8·7	9·9	10·6	8·0
Pro	10·2	4·7	3·8	4·2	5·9
Gly	7·3	8·2	8·8	7·9	8·3
Ala	24·5	11·5	11·8	12·3	15·1
Cys 1/2	0·0	0·0	0·0	0·0	0·0
Val	4·1	6·7	5·8	5·6	7·0
Meth	0·1	0·8	1·2	1·2	0·7
Isoleu	1·2	4·5	5·4	5·4	3·5
Leu	5·0	8·6	8·7	8·9	7·6
φala	0·6	1·3	2·5	2·7	1·4
Tyr	0·7	3·0	2·4	2·3	2·2
Trypt	0·0	0·0	0·0	0·0	0·0

After Rasmussen *et al.*, 1962.

of lysine and arginine are quite different in the different histone components. Fractions I are rich in lysine and poor in arginine. Fractions III and IV are rich in arginine and poor in lysine. Fraction II is intermediate. For this reason it has become fashionable to speak of histone I as lysine-rich histone and fractions III and IV as arginine-rich histone. Fraction II is naturally but clumsily known as slightly lysine-rich histone. This method of description of the separate histone components has the failing that it may blind us to further interesting differences in amino acid composition. Note, for example, that alanine makes up 24 mole per cent of histone fraction I and only 10 to 12 mole per cent of the remaining histones. The sum of glutamic and aspartic acids is also interestingly different in the various histones, increasing as it does from 7 mole per cent in fraction I to 15 mole per cent in fraction IV. And perhaps most important of all, proline constitutes some 10 mole per cent of histone fraction I and decreases steadily to 4 mole per cent in histone fraction IV. As we shall see in the following chapter, these differences in dibasic amino acid and proline content are probably responsible for some of the differences in physical and biological properties which characterize the several histone fractions.

Histone is then composed of four major and separable components. None of these fractions is however homogenous. Each can be further resolved and to this end particularly satisfactory and useful is zone electrophoresis on starch or polyacrilamide gel. By combined column chromatography and starch gel electrophoresis, it has been possible to clearly distinguish histones Ia, Iaa, Ib1, Ib2, IIaa, IIa, IIb1, IIb2, IIIa, IIIb, and IV and to distinguish less clearly a further 6 to 10 minor histone components (Rasmussen *et al.*, 1962). It may be concluded that whole calf thymus histone consists of some 12 major separable components and a further 6 to 10 separable minor entities, 18 to 22 in all.

The heterogeneity of the histones has been pursued in two further ways. The first has been by investigation of the N-terminal amino acids of individual constituents. These studies (Murray, 1964) have revealed that histones of fraction I (both a and b) possess acetyl as the N-terminal group. Histone IIa possesses alanine and IIb proline while fractions III and IV both possess alanine as the N-terminal amino acid. There is then an interesting lack of heterogeneity amongst the histones with regard to N-terminal groups.

The second way we may pursue the heterogeneity question is by tryptic hydrolysis. Trypsin cleaves only arginyl and lysyl bonds and tryptic hydrolysis therefore produces from any protein a finite number of peptides each ending in a lysine or arginine residue. Free lysine and free arginine may be expected too as a result of the cleavage of arginyl-arginyl, lysyl-lysyl, arginyl-lysyl, or lysyl-arginyl sequences. Histone Ia has been studied particularly intensively by tryptic hydrolysis. From its molecular weight, 9500, and its lysine plus arginine content, 28 per cent, it should contain per molecule 29 trypsin-sensitive bonds and should therefore yield 30 peptides in equal molecular amounts. It yields in fact 68 peptides as separated by column chromatography followed by paper chromatography (Murray, 1964). Histone fraction Ia would appear therefore to be a mixture of several kinds of peptide chains, perhaps three or four, all having the same N-terminal group and the same length but differing in sequence. Similarly histone fraction IIb1, expected on the basis of its molecular weight and basic amino acid content to yield approximately 40 peptides, yields in fact 80, while histone fraction IV expected to yield 45 peptides yields 60 and may consist of 2 kinds of peptide chains.

While it is clear that the separation and identification of histones is still a matter in need of much study, present evidence suggests that calf thymus contains some 18 to 22 chromatographically and electrophoretically separable components, and that each of these consists in turn of 2 to 4 or more different kinds of peptide chains. Calf thymus histone is then composed of at least 50 and perhaps even as many as 100 kinds of individual chemical entities. This number, while large, is far smaller than the number of genes in the calf genome. To look at it another way, even though the several kinds of histones yield a substantial number of kinds of peptides on tryptic hydrolysis, this number is by no means as large as the number of peptides which would result from tryptic hydrolysis of a whole calf. Clearly the same histone must occupy and complex with the DNA of several genes.

Next to a second feature of histone structure—the spacing of the basic groups along the peptide chain. It is an interesting fact that histones of fraction I (the lysine-rich) yield on tryptic hydrolysis a substantial proportion of their total lysine content as free lysine. A great deal of their lysine is involved therefore in lysyl-lysyl groupings. These lysine doublets are however spaced at variable distances from one another along the chain since the tryptic peptides of fraction 1 vary in length from 2 to 13 non-basic amino acids. This is true also of the remaining histone fractions in which the spacing of basic amino acids is variable, with 1 to 8 or more non-basic amino acids between each basic group. Basic amino acids doublets involve only about half of the basic residues of histone fraction II and an even smaller fraction of fractions III and IV.

In summary then, the histones are composed of four principle groups. These in turn are further separable into entities which are in turn composed of two or more types of molecules differing in amino acid sequence. In all cases the basic amino acids are scattered at irregular intervals along the peptide chain.

Let us now ask, do different cells, tissues, and organs in the same creature have different or similar kinds of histones? The answer is that broadly the same histones are present in all cells of any given organism. This has been most carefully investigated for calves by Hnilica, Johns, and Butler (1962) who found the thymus spectrum of histones to be present also in spleen and liver. A similar result has been obtained for chickens by Neelin and Butler (1961) who found the same spectrum of histones in spleen, liver, kidney, and heart.

There are however, two major exceptions to the rule that every cell gets the same histones. The first is sperm cells. In the sperm of all animals histones are absent or nearly so, and the genomal DNA is complexed instead with protamine—very short chain polyamines in which two out of three amino acids are either arginine or lysine. The sperm nucleus enters the egg in this condition and only during the course of early cell division is protamine replaced by histone. The second exception is the nucleated erythrocyte of chickens which have been clearly shown by Neelin (1964) to contain substantial quantities of a histone not present as a major component in other cells of chickens. This component, which replaces the usual histones III and IV, is more basic (30 mole per cent) and richer in proline and serine (19 mole per cent) than the histones that it replaces. This curious erythrocyte histone and the fact that it constitutes an apparently real exception to the otherwise uniform distribution of histones through cell types may be associated with the fact that the genome of erythrocyte nuclei is totally or almost totally repressed and will never again be derepressed.

Even though with these few exceptions the histones of the different specialized cells of any organism are in general similar, such cells do differ in one striking way, namely in histone/DNA ratio. We have seen that to fully complex DNA with histone requires a histone/DNA mass ratio of approximately 1·35 to 1. Chromatin in general does not contain histone in an amount sufficient to fully complex all of the genomal DNA. Thus in the developing cotyledon of the pea plant the chromatin contains about 0·9 as much histone as is required for equivalence of histone and DNA. For chromatin of the pea embryo, this number is about 0·8 and for the vegetative bud about the same. Similar observations have been made for maize by Rasch and Woodward (1959). In this plant the amount of histone in nuclei of different cell types varies over the range of 0·8 to 1·0 of that required for full equivalence of histone to DNA. Similarly in the chicken, only in the erythrocyte is histone present in amount equivalent to DNA. In cells of liver spleen and heart the DNA is only about 0·8 complexed. These differences in histone/DNA ratio may reflect differences in the extent to which the genome is repressed. This in turn varies from 100 per cent in the nucleated erythrocyte to a low of perhaps 70–80 per cent. In active cells generally, some 80–90 per cent of the genes still appear to be repressed and only some 10–20 per cent derepressed at any one time. We have come to this conclusion of course, not only from the histone/DNA ratio, but also from consideration of the abilities, relative to pure DNA, of the various kinds of chromatin to support DNA-dependent RNA synthesis in the presence of added RNA polymerase, the matter considered in Chapter 3. Finally, it is clear that the same histone must repress different genes in different tissues.

We have thus far considered the histones of the different cells of any single creature. Let us now consider the histones of different creatures. The principal fact is that not only are their amino acid compositions very similar, but so also is the characteristic manner in which they may be separated into four principal sub-groups. Histones Ib for example are found in pea, rat, rice, and *Chlorella* as well as in thymus, and in addition, these histones have similar chromatographic and electrophoretic properties, and similar amino acid compositions (Bonner and Ts'o, 1964). The same is true for the other principal histone components. It is true that histones of plants, some tissues of the pea, wheat germ, rice, and *Chlorella* are somewhat impoverished in fractions III and IV as compared to histones of thymus—a fact which is now elevated to the law that plant histones are richer in lysine than are animal histones. But this is the principal present exception and it is merely a quantitative one at that. The more general principle is that the creatures thus far studied possess similar histones.

Nor are the similarities between the histones of different creatures confined to composition and physical properties. They extend also to amino acid sequence. One of the more careful investigations of this matter is that of Hnilica *et al.* (1964) who have compared the tryptic peptides of calf thymus histones IIb with those of histone IIb of a rat tumour, the Walker tumour. The data of Table 2 compare the compositions of the corresponding peptides in the two cases. Of the nine peptides considered six are identical. The remaining three differ as between calf and rat, by one amino acid residue each. This reminds us of the differences between corresponding enzymes of different creatures; their haemoglobins or their insulins. These differences in amino acid composition and sequence are attributable to mutation in the genes involved. In the case of the histones we imagine that they too are genetically controlled and that over the millenia different mutations have accumulated in different creatures although not in sufficient number to becloud the similarities of the proteins. We may conclude that the control of gene structure and function requires particular histones of particular structure and this independently of the gene, and the creature—rat, cow, or pea—to be controlled.

Histones are synthesized during the interphase between mitotic cell divisions as is DNA, and the newly synthesized DNA is quickly enshrouded in appropriate histone. So synchronous are DNA and histone synthesis that it at once occurs to one that the two processes may be linked to one another, and this suggestion has in fact been often made. To discover whether the two processes are linked we have tried first to find whether the two processes can be uncoupled. They can. To this end Flamm and Birnstiel (1964) have harnessed the nuclear technology discussed in Chapter 7. They grow dissociated cells in tissue culture in such a manner that they grow exponentially with a doubling time of two days. In such cells, DNA and histone synthesis normally parallel one another. DNA synthesis can however be suppressed with the inhibitor 5-flurodeoxyuridine (5-FDU), which specifically abolishes the

TABLE 2

Comparison of tryptic peptides of histone fraction IIb *prepared from calf thymus with those from the same histone fraction prepared from rat tumour* (*Walker*)

Peptide No.	Source	Amino acids present in peptide
1	Calf thymus	(GLU, GLY, SER$_3$, TYR, VAL) LYS
	Rat tumour	(GLU, GLY?, SER$_3$, TYR, VAL) LYS
2	Calf thymus	(ASP, GLU, GLY, HIS, ILE, PRO, SER$_2$, THR, VAL) LYS
	Rat tumour	(ASP, GLY, GLY, HIS, ILE, PRO, SER$_2$, THR, VAL) LYS
3	Calf thymus	(ASP, GLY) LYS
	Rat tumour	(ASP, GLY) LYS
4	Calf thymus	(ALA, GLU, ILE, THR, VAL) ARG
	Rat tumour	(ALA, GLU$_2$, ILE, THR, VAL) ARG
5	Calf tumour	(ALA, GLU, GLY, LEU$_3$, PRO) LYS
	Rat tumour	(ALA, GLU, GLY, LEU$_4$, PRO) LYS
10	Calf thymus	(ALA, LEU, TYR,) HIS
	Rat tumour	(ALA, LEU) HIS
14	Calf thymus	(ALA, THR, VAL) LYS
	Rat tumour	(ALA, THR, VAL) LYS
15	Calf thymus	(TYR, VAL) LYS
	Rat tumour	(TYR, VAL) LYS
16	Calf thymus	(LEU, VAL) LYS
	Rat tumour	(LEU, VAL) LYS

After Hnilica *et al.*, 1964.

formation of thymidylate. 5-FDU in a concentration of 10^{-6} M achieves this end without inhibiting RNA synthesis. To such cells in which DNA synthesis has been suppressed, Flamm and Birnstiel now supply labelled amino acid, lysine, or leucine. After protein synthesis has taken place in the presence of the labelled amino acid for a short time, say 1 hour, the cells are harvested, their nuclei isolated, and from them histone is extracted. Histone synthesis continues unabated for at least 18 hours after the cessation of DNA synthesis. Synthesis of histone does not then require the presence of a growing DNA chain. The two processes are wholly dissociable.

Where in the nucleus does histone synthesis take place? It appears the histones are synthesized within the nucleolus. This is indicated by two facts, namely:

(1) When cells actively engaged in histone synthesis are incubated in labelled amino acid for long periods the bulk of the label incorporated into histone is found of course in the chromatin and only a small amount in the nucleolus. As the period of incubation is progressively shortened we arrive ultimately at an incubation period sufficiently brief, 20 seconds, so that histones become labelled in the nucleolus alone (Flamm and Birnstiel, 1964).

(2) Isolated nucleoli but not isolated chromatin can synthesize histone (Flamm and Birnstiel, 1964). We conclude therefore that histone synthesis is one function of the nucleolus. Once synthesized, histones must be, in some manner transported to and complexed with the newly formed DNA of the replicating genome. Of these operations we have as yet no knowledge.

The synthesis of histones occurs by the familiar ribosomal decoding of messenger RNA system, since it is inhibited both by puromycin and by actinomycin-D. Interestingly enough, the messenger RNA for histone synthesis appears to turn over very rapidly in the nucleus. Thus Schmeiger (1964) has shown that 15 minutes after the application of actinomycin-D to ascites cells, the rate of histone synthesis has decayed to one half its initial value. The decay rate of the messenger RNA for histone synthesis is greater than for any other class of nuclear or cytoplasmic protein according to Honig and Rabinowitz (1964). It is already known that a part of the nuclear RNA not only turns over rapidly (Harris et al., 1963; Hiatt, 1962) but is degraded in the nucleus at a rate which is similar to that of the messenger RNA of histone synthesis. It is tempting to associate the two.

Now to two final facts concerning the histone economy of the nucleus. We all know that DNA segregates semiconservatively at mitosis. Each strand of the original double helix remains intact through DNA replication and is ultimately partitioned at mitosis to a daughter cell. The histone of the chromatin is not however partitioned in this semiconservative fashion as has been shown by Prescott (1964). The autoradiographic experiments of Prescott are exactly analogous to those used to demonstrate semiconservative segregation of DNA. In Prescott's experiments, cells of the Chinese hamster were supplied with radioactive amino acid during intermitotic chromosome doubling. The label was then removed and the segregation of protein in the labelled chromosomes followed. At the mitosis following labelling, the histone was shared between the daughter chromosomes and thus passed to each of the daughter cells. At the second mitosis, that at which labelled strands of DNA segregate, histone was on the contrary again shared between daughter chromosomes. It is concluded that histone once deposited on the DNA molecule does not stay there in perpetuity but can come off and be re-assorted during chromosomal multiplication.

The DNA of resting cells is of course metabolically inert—it is neither synthesized nor degraded. DNA does not turn over; it is immortal. What can we say of the histone component of the chromatin of resting cells. The histone, unlike the DNA is not immortal. In resting cells, histone remains constant in amount but is slowly degraded and replaced by freshly synthesized material. In the liver for example this rate of turnover is approximately 5 per cent of total histone per 24 hours according to Busch and Steele (1964). This slow turn-over of histone suggests interesting possibilities concerning the feasibility of gene induction and derepression even in non-dividing cells.

We have seen then that nature provides the genome with a small multiplicity of histones. This could not have been anticipated. One line of reasoning would suggest that one kind of histone, a gene-turning-off kind, should suffice, if the role of histones is to turn off genes. A second line of reasoning might suggest that since there are many different kinds of genes to be repressed an equal number of species of histones should be required. But neither is correct. There are a few kinds of histones: these are not synthesized on the genes to be repressed but are synthesized in a central depot, the nucleolus. Why

7

does the cell provide itself with a small multiplicity of histones? To answer this question we shall need to enquire into the biological properties of the individual nucleohistones. This subject shall be considered in the following chapter.

SELECTED REFERENCES

General summary of histone chemistry

BONNER, J. and TS'O, P. O. P. (Editors), *The Nucleohistones*. Holden-Day, San Francisco (1964).

PHILLIPS, D. M. P., *Prog. Biophys. biophys. Chem.* **12**, 211 (1962).

Speculation as to histone function

STEDMAN, E. and STEDMAN, E., *Phil. Trans.* **B235**, 565 (1951).

Histone separation

RASSMUSSEN, P. S., MURRAY, K. and LUCK, J. B., *Biochemistry* **1**, 79 (1962).

JOHNS, E. W. and BUTLER, J. A. V., *Biochem. J.* **82**, 15 (1962).

Tryptic peptide analysis of histone

MURRAY, K., in *The Nucleohistones*.

JOHNS, E. W., in *The Nucleohistones*.

PHILLIPS, D. M. P., in *The Nucleohistones*.

HNILICA, L. S., TAYLOR, C. W. and BUSCH, H., in *The Nucleohistones*.

BUSCH, H., STEELE, W. J., HNILICA, L. S. and TAYLOR, C. W., in *The Nucleohistones*.

Comparison of histones of different cell types

HNILICA, L. S., JOHNS, E. W. and BUTLER, J. A. V., *Biochem. J.* **82**, 123 (1962).

NEELIN, J. M. and BUTLER, J. A. V., *Can. J. Biochem. Physiol.* **39**, 485 (1961).

NEELIN, J. M., in *The Nucleohistones*.

RASCH, E. and WOODWARD, J., *J. Biophys biochem. Cytol.* **6**, 263 (1959).

Site and properties of histone synthesis

FLAMM, W. G. and BIRNSTIEL, M., in *The Nucleohistones*.

HONIG, G. R. and RABINOWITZ, M., *Fed. Proc.* **23**, 268 (1964).

SCHMEIGER, H. G., *Fed. Proc.* **23**, 382 (1964).

Rapidly turning over nuclear RNA

HARRIS, H., *Proc. Roy. Soc.* **B157,** 177 (1963).
HIATT, *J. mol. Biol.* **5,** 217 (1962).

Segregation of histone

PRESCOTT, D., *The Nucleohistones*

Turnover of histones

BUSCH, H. and STEELE, W. J., in *The Nucleohistones.*

THE ANNEALING OF HISTONE TO DNA

WE HAVE seen that the histone component of chromatin is a mixture of a rather small number of kinds of basic proteins and that they may be grouped into four major classes. These classes are based upon differences in chromatographic and electrophoretic behaviour which in turn are attributable to different amino acid compositions. We approach the question of why nature provides these various histones by study of the interaction between DNA and histone and of the properties of the nucleohistones thus generated.

If we mix histone and DNA in medium of low ionic strength, histone binds randomly to DNA, cross-linking the individual DNA molecules into a gel, which precipitates from solution as a horrid glob. Much work on nucleohistone reconstitution stops at this point. We can, however, approach the matter more subtly. Suppose we mix histone and DNA in a solution of salt concentration sufficiently high so that both materials are soluble. In this medium, by virtue of its high ionic strength, histone and DNA do not bind to one another. Suppose we next dialyze against progressively lower salt concentrations until we reach a concentration such that histone-DNA interaction becomes appreciable. Such a concentration is 0·4 M sodium chloride. At this concentration histone binds lightly to DNA cross-linking DNA molecules so that slight but detectable gel formation takes place. Suppose we now let the system remain at this concentration. Histone molecules pop off and on the DNA and, with time find, and remain in conformations of maximum stability. These conformations are evidently ones in which individual DNA molecules are covered with histone molecules and without cross-links. This is evidenced by the fact that if the salt concentration is lowered still further, say to distilled water, no precipitation of cross-linked nucleohistone takes place. This is an interesting fact. It indicates that there is something special and specific about the interaction between histone and DNA. There is some aspect to this interaction which confers upon it a preferred geometry. Perhaps it is due merely to the geometries of DNA and histone. Perhaps there are non-ionic forces at work such as stacking interactions between histone side chains and DNA—forces which are weak compared to ionic ones, but which can still make one nucleohistone configuration more stable than another.

In any case it is clear that the mode of preparation of soluble nucleohistone is essentially annealing. It is analogous to the annealing of renatured DNA. In this case, denatured DNA consisting of a mixture of the complementary strands from many different DNA molecules is left under appropriate conditions, and each strand ultimately finds and pairs with its complement. Our annealing of nucleohistone is similar too, to the annealing in which a given messenger RNA finds and pairs with that gene, which alone of all the genes of the genome, produced it. In all these instances we use conditions in which the difference between the binding energies of the more stable and the less stable complex is maximum. The annealing of nucleic acid chains is achieved by the use of temperatures just high enough to permit DNA molecules to bind loosely, test one another for fit, and pop off and seek another if they do not fit. They are conditions which permit of hunting. With our nucleohistone in which the interaction is primarily ionic, we use an ionic strength which permits such hunting.

We now have reconstituted nucleohistone. Like the native nucleohistone preparation from chromatin, of which we have spoken earlier, our reconstituted preparation is optically clear at wavelengths greater than 305 mμ—the solutions are not turbid. The spectrum of reconstituted nucleohistone is distinguished from that of DNA most strikingly by its greater absorption at 230 mμ due to the peptide bonds of the histone (Fig. 1). In our nucleohistone preparation histone is present in an amount equivalent to DNA, an amount in which basic groups of histone just equal phosphate groups of the DNA. The histone: DNA mass ratios of the reconstituted nucleohistones are therefore in the region of 1·35 to 1. The data of Table 1 show that in such a reconstituted nucleohistone, histone is in fact bound to DNA. For the experiment of Table 1,

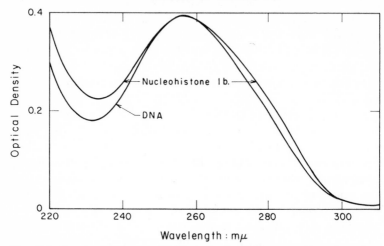

Absorption Spectrum of Thymus DNA as
Influenced by Binding with an Equivalent
Amount of Histone I b.

FIG. 9.1. Spectra of thymus DNA and of soluble reconstituted nucleohistone prepared by annealing a stoichiometric amount of thymus histone
Ib to thymus DNA. After Huang *et al.* (1964).

TABLE 1

*Demonstration that the histone component of reconstituted
nucleohistone accompanies the* DNA *under conditions of centrifugation suitable for sedimentation of the* DNA *but not of histone
alone. Centrifugation for* 21 *hours at* 130000 × g. *Nucleohistone
and* DNA (*thymus*) *dissolved in dilute saline citrate. After
centrifugation, amounts of* DNA *and of nucleohistone in supernatant and pellet analyzed by* O.D. *at* 230 *and* 260 mμ

| Nucleohistones | Histone/DNA mass ratio† | Distribution of material† after centrifugation | |
		Supernatant (%)	Pellet (%)
DNA	—	2·2	98
Nucleohistone Ib	1·37	1·3	99
Nucleohistone IIb	1·32	0·5	99
Nucleohistone III	1·45	0	100
Nucleohistone IV	1·35	2·7	97

† Histone/DNA ratio of pellet material was identical with that of initial material,
i.e. nucleohistone.   After Huang *et al.* (1964).

samples of each of a series of nucleohistone solutions were centrifuged for 21 hours at $130000 \times g$, a condition which completely sediments DNA, but which, as shown by separate experiments, does not sediment any molecularly dispersed histone. Under these conditions the histone accompanies DNA to the pellet.

The presence of histone in nucleohistone also influences the sedimentation behaviour of the DNA itself. The data of Table 2

TABLE 2

Sedimentation behaviour of thymus DNA *and of nucleohistone reconstituted from it with histone fraction* IIb

	Thymus DNA	Nucleohistone IIb
% of total material not sedimented at $24000 \times g$, 20 min	100	83
% of total material not sedimented at $130000 \times g$, 15 min	100	43
Sedimentation coefficient of material not sedimented at 130000 $\times g$, 15 min	$17 \cdot 3\ S_{20}$	$26 \cdot 2\ S_{20}$

After Huang *et al.*, 1964.

show that the sedimentation coefficient of a sample of DNA is raised from 17 S to 26 S units by full complexing with histone, exactly the increase expected if the mass of the DNA molecule is increased by a factor of $2 \cdot 4$ as it would be by binding an equivalent amount of histone. This relation between the sedimentation coefficients of DNA and of nucleohistone, is too, exactly that found by Giannoni and Peacocke (1963) between the sedimentation coefficients of native chromosomal nucleohistone and the DNA derived from it.

We have seen earlier that in chromosomal nucleohistone, the histone stabilizes the DNA against melting. Let us then enquire as to the effectiveness of the several histone fractions in such stabilization of DNA. Figure 2 presents data on the melting profile of thymus DNA and of the nucleohistone made therefrom by binding with histone Ib. It is clear that histone Ib does appreciably stabilize DNA against melting. It also causes the melting profile to be more diffuse than it is for DNA itself. We take this fact to indicate that histone Ib binds to DNA in a

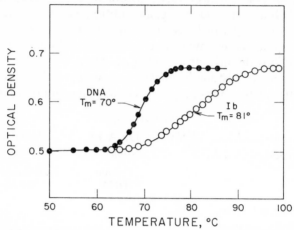

FIG. 9.2. Melting profiles of thymus DNA and of soluble reconstituted nucleohistone Ib. Melted in dilute saline citrate, 0·016 M. After Huang *et al.* (1964).

variety of conformations each contributing a different degree of stabilization.

Stabilization of DNA against melting is not a property of histone IV as is shown in Fig. 3, even though, as seen in Table 1,

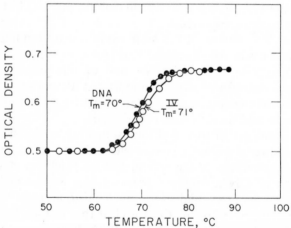

FIG. 9.3. Melting profiles of thymus DNA and of soluble reconstituted nucleohistone IV. Melted in dilute saline citrate, 0·016 M. After Huang *et al.* (1964).

histone IV is in fact bound to, and sediments with, DNA. The data of Table 3 summarizes information concerning the melting behaviour of the several nucleohistones. It is apparent that these form a regular series, histone Ib contributing the greatest stabilization to DNA, histone IV the least, while histones IIb and III are intermediate. The properties of nucleohistone reconstituted from unfractionated thymus histone are as seen

TABLE 3

Histone/DNA ratios and half melting temperatures, T_m, of reconstituted nucleohistones. Melting done in dilute saline citrate, 0·015 M NaCl, 0·001 M Na citrate

DNA as	Histone/DNA mass ratio in soluble nucleohistone	T_m of nucleohistone (°C)
Nucleohistone Ib	1·37	81
Nucleohistone IIb	1·32	75
Nucleohistone III	1·45	72·5
Nucleohistone IV	1·35	71
Whole thymus nucleohistone	1·33	76
Nucleoprotamine (salmine)	—	70
DNA alone	—	70

After Huang *et al.*, 1964.

from the data of Table 3, intermediate between those of nucleohistones Ib and IIb. This is as expected, since as pointed out in Chapter 8, histones Ib and IIb are the principal components of calf thymus histone.

The data of Table 3 include information concerning nucleoprotamine in which DNA is fully complexed with salmine, the protamine of salmon sperm. Salmine, a peptide approximately twenty amino acid residues in length and in which two out of every three residues are arginine, does not stabilize DNA against melting.

What determines the degree to which different histones and protamines stabilize the structure of DNA? It does not appear that this property is one which determines the firmness with which a given histone is bound to DNA. The equilibrium constants for the binding of DNA to the histones and to protamine, have been determined by the rigorous method of equilibrium dialysis by Akinrimisi *et al.* (1964). The affinity of fraction Ib for DNA is actually slightly less than that of fraction IV. The differences in DNA-stabilizing ability more probably reside in differences in the structures of the nucleohistones themselves, differences due to differences in the secondary structure of the histones. Histone Ib, because of its high proline content, cannot assume alpha helical conformation. In addition, because of the average of six amino acid residues between lysine doublets in the peptide chain it is not easy for the molecule to bind successive phosphate residues in the same DNA strand. We can therefore imagine that the histone Ib chain coils around the DNA molecules, binding now two phosphate groups in one DNA strand, now two phosphate groups in the second strand and as a consequence, binds the two strands together. Histones III and IV can assume a considerable degree of alpha helical conformation. Bradbury *et al.* (1964) and Zubay and Doty (1959) have shown that the whole histone of native nucleo-histone contains an appreciable amount of alpha helical structure. In such a molecule basic residues will protrude, on the average, one per turn of the alpha helix and will thus be relatively closely spaced, a structure which would permit individual histone molecules to bind individual DNA strands along their length.

But this is speculation; we do not in fact know the structures of the several nucleohistones, nor even that of native nucleo-histone itself. The X-ray diffraction patterns of nucleohistones consist principally of diffuse rings and poorly orientated arcs, the patterns being much less sharp than those of DNA alone. The X-ray investigations do reveal that the structure of DNA is preserved in native nucleohistone. For example the meridional spacing of 3·4 Å due to the stacking of the base pairs along the helix axis is clearly evident as are the layer-lines characteristic of the DNA double helix. A meridional spacing of about 110 Å may be due to super-coiling of the nucleohistone molecules. One principal reason for the lack of progress in the study of the structure of nucleohistones is no doubt the fact that only nucleohistone containing all the different histones of thymus have been investigated. As we have seen, the different individual histones probably contribute different structures to the nucleo-histone ensemble. It will be therefore of great importance to our understanding of nucleohistone matters, to investigate by X-ray diffraction the structure of nucleohistones reconstituted from individual histone components.

Let us now consider the abilities of the reconstituted nucleo-histones to support DNA-dependent RNA synthesis by the purified chromosomal RNA polymerase of pea-embryo chro-matin. For these experiments the solubilized chromosomal

TABLE 4

Activities of various reconstituted nucleohistones in support of RNA *synthesis by E. coli* RNA *polymerase*

DNA (50 μg) provided as:	RNA synthesized $\mu\mu$m nucleotide/0·5 ml/10 min†
DNA (thymus) alone	8474
Nucleohistone Ib	56
Nucleohistone IIb	140
Nucleohistone IV	4000
Nucleoprotamine (salmine)	7287

† Incorporation by enzyme alone (8 $\mu\mu$m) subtracted.
After Huang *et al.*, 1964.

RNA polymerase is incubated in a reaction mixture containing the four riboside triphosphates, one of them C^{14}-labelled, and DNA, either as nucleohistone or as pure deproteinized material. Deproteinized pea DNA of course supports DNA-dependent RNA synthesis by the pea-embryo chromosomal polymerase. The various reconstituted nucleohistones show a graduation in their effectiveness in support of DNA-dependent RNA syn-thesis. This graduation parallels the modifying effect of the various histones on the thermal denaturation of DNA. Thus the nucleohistone formed with histone Ib is completely inactive in supporting RNA synthesis, and nucleohistone IIb almost so. Nucleohistones III and IV however support RNA synthesis to a considerable extent. The polymerase of *E. coli* (Table 4) like that of pea-embryo chromatin is unable to synthesize appreciable RNA in the presence of nucleohistone Ib. Nucleohistone IIb is also barely active in the support of RNA synthesis. Nucleo-histones III and IV are highly active in the support of RNA synthesis and nucleoprotamine is nearly as effective as de-proteinized DNA itself. With both polymerases, nucleohistone reconstituted from DNA and the unfractionated mixture of thymus histone is essentially inactive in the support of RNA synthesis. This again is as might be expected since the whole histone consists principally of histones Ib and IIb.

Addition of histone to DNA in solution of low ionic strength causes, as noted above, immediate aggregation of the DNA.

TABLE 5

Influence of varied histones on the capacity of DNA to support RNA synthesis by RNA polymerase. Histone, in amount equivalent to DNA, added directly to reaction mixture

Addition to otherwise complete reaction mixture	Rate of RNA synthesis $\mu\mu$m nucleotide/0·5 ml/10 min
DNA (no histone)	5340
DNA + histone Ib	450
DNA + histone IIb	850
DNA + histone III	3290
DNA + histone IV	3770

After Huang *et al.*, 1964.

The aggregates thus formed by this instant reconstitution do possess however, some of the attributes of the soluble nucleohistone prepared by the annealing process. The data of Table 5, concerning the influence of added histone and of protamine upon rate of RNA synthesis by *E. coli* RNA polymerase. To the complete reaction mixture an amount of histone equivalent to the DNA was added immediately before addition of the polymerase. The abilities of these aggregates to support RNA synthesis parallel those of the soluble reconstituted nucleohistones. It is particularly striking that the aggregate produced by histone IV and DNA is essentially as effective as DNA itself in the support of RNA synthesis. Aggregation and precipitation from solution does not of itself impare the ability of DNA to support DNA dependent RNA synthesis.

It is apparent then, that the degree of inactivity of a reconstituted nucleohistone in supporting DNA-dependent RNA synthesis interestingly parallels the extent to which the histone fraction in question stabilizes DNA against thermal denaturation. It may be supposed that the differences between the several nucleohistones in both of these properties are due to differences in structure. Again such differences may be associated with differences in alpha helical content of the histones.

Let us now ask, how are the several kinds of histones partitioned amongst the DNA molecules of the genome? Does each DNA molecule receive a complete and representative sample of the various histone molecules, or does each DNA molecule get only one kind of histone? We approach this matter through the electrophoretic properties of the nucleohistones. Such studies have been briefly reported by Davidson (1964). Free boundary electrophoresis through a medium stabilized by a sucrose density gradient has been found suitable for such studies.

DNA has a characteristic electrophoretic mobility due to its high negative charge density. All nucleohistones too are negatively charged, for although the cationic groups of histone equal the phosphate groups of DNA, histones possess in addition the carboxylate ions of glutamic and aspartic acids. But, as we have seen, the four principal histone fractions differ greatly in their content of these groups. The anionic charge densities of the several reconstituted nucleohistones vary then, from slight in nucleohistone Ib to considerable in nucleohistone IV, and their mobilities from slight for nucleohistone Ib to about three fourths that of DNA for nucleohistone IV. Each reconstituted nucleohistone migrates therefore with its characteristic mobility. In addition, their migration is little affected by the presence of DNA or of another nucleohistone, so that mixtures are nicely resolvable into their components.

Let us now reconstitute a nucleohistone from the whole mixture of histones. We might hope that each DNA molecule would receive a random mixture of all histones present in the mixture. This hope is justified because the reconstituted whole thymus nucleohistone migrates as a single band of mobility close to that of its major component, nucleohistone II. We can now answer the question set above, namely, how are the histones partitioned amongst the DNA molecules in the native nucleohistone of chromatin? To answer this question we electrophorese some soluble native nucleohistone. It does not migrate as a single component as did the reconstituted material. Native nucleohistone is in fact electrophoretically very heterogenous, exhibiting mobilities as slight as those due to components of Ib and IIb as well as a broad front of more rapidly moving material of mobilities resembling those of nucleohistones III and IV. This suggests that in chromatin particular DNA molecules are complexed principally with histone of one particular type. Each DNA molecule does not get a random assortment but rather a preferred category of histone. Evidently native nucleohistone does not contain DNA molecules complexed solely with histone of type IV. Such complexes should support RNA synthesis, and we know that native nucleohistone is not active in this function. But why then are there histones such as III and IV in chromatin? Do they function as linkers, linking the nucleohistone strands side by side? Do they have to do perhaps with chromosome coiling, or do they have perhaps to do with genes which are particularly susceptable to derepression? We do not yet know, but our chromosomal technology will now make possible the study of questions such as these.

Let us phrase our question more broadly. Why does chromatin contain different but well defined kinds, classes, of histones? One partial answer is that suggested above—that some kinds of histones may have to do with chromosomal structure rather than with the control of chromosomal activity. But with regard to chromosomal activity itself, in the higher creature, if it is to be a proper higher creature, one and the same gene must possess different attributes, different attitudes, in different cells. Most simply a gene may be turned on in one cell, repressed in another. This is the way in which we have thus far discussed the situation. But this is not all there is to it, for repression must take many forms. A gene in one cell may be repressed but competent to be derepressed by a particular signal, while the same gene in another cell is not only repressed, but is also not competent to be derepressed by the same signal. Take for example the induction of flowering about which much is known. A signal is generated in leaves in response to an environmental stimulus. This signal travels out into the plant body and says 'Genes for flowering awake! Become derepressed! I have seen a short day and it is therefore time for you to make your messenger RNA's and hence the enzymes which are required for the making of our flowers'. Fine, but of all the cells in the plant only a few respond to this message, only those of the growing buds. The very same genes in the growing roots are repressed too, but they stay repressed; roots do not flower. This is what we understand by differences in gene attitude and it is to provide these differences in gene attitude, we hypothesize, that nature provides a multiplicity of histones. According to this view, one kind of histone may render a gene repressed and not derepressible by any signal. A second kind of histone may render the same gene repressed but depressible by signals of type one, a third kind of histone may render the gene repressed but derepressible by signals of type two, and so on. According to this view, the partition of the several kinds of histone amongst the DNA molecules of the genome reflects the strategy of the programming of gene activity to which development is due.

This is however merely an hypothesis, a working hypothesis intended as a guide to experimentation. What kinds of experiments does it suggest? One is to fractionate the genome by electrophoresis, focusing on a particular gene, and to find whether this same gene is complexed with different histones in different cell types. This is logistically difficult, but logically possible. Another experiment suggested by our hypothesis is to look into the matter of the signals which evoke gene activity. Do different signals in fact operate on different species of nucleohistone? It is with this matter, the internal signals which evoke gene activity, that we shall concern ourselves in Chapter 11.

SELECTED REFERENCES

Histones and the transcription of DNA

HUANG, R. C. C. and BONNER, J., *Proc. Nat. Acad. Sci., Wash.* **48**, 1216 (1962).

BARR, G. C. and BUTLER, J. A. V., *Nature, Lond.* **199**, 1170 (1963).

HINDLEY, J., *Biochem. biophys. Res. Commun.* **12**, 175 (1963).

Annealing of histones to DNA

HUANG, R. C. C., BONNER, J. and MURRAY, K., *J. mol. Biol.* **8**, 54 (1964).

Physical properties of nucleohistones

ZUBAY, G. and DOTY, P., *J. mol. Biol.* **1**, 1 (1959).

GIANNONI, G. P. and PEACOCKE, A. R., *Biochem. biophys. Acta* **68**, 157 (1963).

HUANG, R. C. C., BONNER, J. and MURRAY, K., *J. mol. Biol.* **8**, 54 (1964).

AKINRIMISI, E. O. TS'O, P. O. P. and BONNER, J., *J. mol. Biol.* In press.

DAVIDSON, N., in *The Nucleohistones*. Holden-Day, San Francisco (1964).

Studies of nucleohistone structure

ZUBAY, G. and DOTY, P., *J. mol. Biol.* **1**, 1 (1959).

WILKINS, M. H. F. ZUBAY, G. WILSON, H. R., *J. mol. Biol.* **1**, 179 (1959).

ZUBAY, G. and WILKINS, M. H. F., *J. mol. Biol.* **4**, 444 (1962).

HUXLEY, H. E. and ZUBAY, G., *J. Biophys. biochem. Cytol.* **11**, 273 (1961).

BRADBURY, E. M., PRICE, W. C., WILKINSON, G. R. and ZUBAY, G., *J. mol. Biol.* **4**, 50 (1962).

RICHARDS, B., in *The Nucleohistones*.

BRADBURY, E. M., in *The Nucleohistones*.

10

REGULATOR, OPERATOR, EFFECTOR

WE HAVE so far discussed the control of genetic activity only in terms of what has been done by isolation of chromosomes with their control machinery intact. There are other quite different ways to study genetic control mechanisms, however. Let us survey the world of genetic control. Firstly, there is a small and scattered group which works upon the programming of genetic information in bacteriophage T4. This work has shown that even viruses have genetic control although we do not know how it works. Next there are those who study genetic control of genetic activity works in *E. coli*. This approach is exemplified by Jacques Monod, Francois Jacob, and their allies of the Pasteur Institute, Paris. At a further level there are those who investigate the genetics of genetic control in creatures slightly more complicated than bacteria such as *Neurospora* and *Aspergillus*, exemplified by N. H. Horowitz and G. Pontecorvo.

We will now take cognizance of this other world, the genetic approach to the genetic control world, and see how the other half lives. What have they found out and how does it compare with, and contribute to, what we have learned from the study of chromosomes? We begin with the study by Jacob and Monod of the production of a particular enzyme of *E. coli*. The story is a familiar one. Nonetheless, by going through it once more with fresh eyes we may see something new and relevant to our problem.

The gene of *E. coli*, upon which we focus our attention, is that which controls the production of β-galactosidase. β-galactosidase is an inducible enzyme. It is not made by *E. coli* cells (of appropriate constitution) in any substantial quantity until a β-galactoside comes along to be cleaved. In the presence of its substrate, the enzyme is produced, made *de novo* from amino acids. The enzyme is therefore not stored in some inactive form prior to induction. The production of the enzyme requires the synthesis of appropriate messenger RNA. A full account of the kinetics of each step in the production of β-galactosidase in *E. coli* has been provided by Kepes (1963) and for a second inducible enzyme by Hartwell and Magasanik (1963). But the production of the enzyme itself turns out to be nothing but classical molecular biology. The β-galactosidase gene is transcribed by RNA polymerase, the appropriate messenger RNA is made and this in turn is decoded by a ribosome with production of the enzyme. We focus our attention not upon the chemistry but upon the genetics of the process.

There are three types of mutations which influence the production of β-galactosidase by *E. coli* cells. The first are those at locus Z within the genome. These lead to the production of no normal enzyme. β-galactosidase can simply not be produced under any circumstances by an *E. coli* cell containing an appropriate mutation at this locus. The Z locus is responsible for producing the messenger RNA which in turn produces β-galactosidase. A mutation in this locus leads to faulty messenger RNA and hence to the altered or non-existent enzyme. We say that the Z locus is the structural gene for β-galactosidase. As a matter of fact it consists of at least two genes side by side, one for the production of β-galactosidase itself and a second for the production of a particular enzyme, β-galactoside permease which catalyzes the entry of β-galactoside into the cell.

The second group of mutations which affect β-galactosidase production are those of the locus which we will call R. Mutant R cells possess β-galactosidase as a normal constituative enzyme. The enzyme need not be induced; the inducer need not be present. β-galactosidase is poured out under all circumstances. The R locus is in a different region of the genome, from the structural gene, Z. Furthermore, the normal unmutated R locus can exert its effect upon a Z locus at a substantial distance. For example in cells into which an extra chromosome has been introduced, an R locus in one chromosome can affect a Z locus in the second. Thus in the diploid, R^+Z^-/R^-Z^+ β-galactosidase is inducible. We must therefore conclude that the gene R makes something which goes to the structural gene Z and regulates its ability to make messenger RNA. The R locus is thus the regulator gene for the β-galactosidase gene. Exactly the same kind of regulation is at work with many genes of *E. coli*—possibly with all. Take the tryptophane synthesizing system for example, in which a single R locus regulates a whole series of genes involved in tryptophane synthesis. In this case the genes in question lie side by side down the genome. Or more spectacularly, consider the genes for the enzymes of arginine synthesis. A single regulator locus controls many structural genes scattered through the genome.

Two general principles emerge, namely:

(1) that in an inducible system the regulator gene produces something which inhibits the structural gene. When an appropriate metabolite comes along it interacts with this something and renders it inactive and unable to repress.

(2) that in a repressible system, the regulator gene produces something which does not inhibit the structural gene until this something has interacted with the metabolite. Thus in the tryptophane and arginine systems, the complex between the metabolite and the product of the regulator represses the structural genes for the enzymes which produce that metabolite.

A single regulator gene can regulate many structural genes. How can the product of the regulator gene recognize these various genes as related? We say that the genes have a common operator. The operator is in a sense a prefix to the structural gene and describes the group to which the gene in question belongs. This leads us to the third class of mutations which affects the production of β-galactosidase. Mutations of the operator gene, 0, are recognized because they prevent the structural gene in question from being repressed even though the regulator locus is intact. Operator mutations are dominant; for example, in the diploid $R^+O^+Z^+/R^+O^-Z^+$ the cell makes the product of Z, β-galactosidase, even in the absence of inducer. No product of the unmutated O locus can move from one chromosome to a second to there operate upon the structural locus. The O locus functions only along the chromosome which contains it. Further, the O locus is always immediately adjacent to the structural gene it controls. One operator may control more than one structural gene so long as these genes are adjacent to one another.

An operator gene together with its adjacent structural genes is known as an operon. We picture the arrangement as outlined in Fig. 1. In this formal diagram regulator produces something which acts upon operator and the operators in turn act upon the adjacent structural gene or genes. With an inducible enzyme the something produced by regulator works upon operator only in the absence of inducer. In the presence of inducer, regulator substance and inducer interact to produce a non-repressor, a material without repressor activity. With a metabolite-repressible gene, the product of regulator interacts with

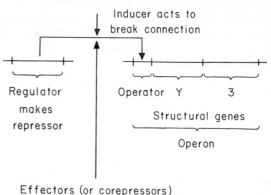

FIG. 10.1. Inter-relationship of regulator, operator, and structural genes. The regulator makes a repressor which by interacting with the operator represses the structural genes. This repressor can be either (1) ineffective in the presence of a particular small molecule (inducible system) or (2) effective only in the presence of a particular small molecule (repressible system).

and inhibits the structural gene only if the product of regulator is combined with the appropriate effector metabolite. Finally we should remember that products of a single regulator can move to, and combine with, the operators of one, two, three, or many operons. With β-galactosidase a single regulator regulates a single operon. With the arginine synthesis system a single regulator regulates many operons.

These then are facts concerning regulation which have been deduced from genetic machinations. The facts are impressive, the logic irrefutable, and the experiments elegant to a high degree. There are also facts which suggest the existence in higher creatures of genes which control other genes. One of these, the first to be described, is due to Rhodes (1941). A genetic locus in maize when mutated, causes occasional springing into activity of genes not linked to that which had been mutated. A further case is that worked out in detail by McClintock (1956). A particular locus, again in maize, when mutated, causes derepression in endosperm of genes for antho-cyanin synthesis. We understand in still greater detail the case in Oenothera studied by Lewis (1960). He worked with the system which controls the growth of pollen tubes in the style and has clearly shown that two genes are involved. One is a structural gene responsible for making an enzyme which makes a specific substance. The second is a control locus for the first. In creatures less complex than higher plants and animals but more complex than bacteria, regulator genes for structural genes are also known, as in *Neurospora* (Horowitz *et al.*, 1961), yeast (Halvorson, 1961), and *Aspergillus* (Pontecorvo, 1963).

Even though there is not, in higher creatures, any case worked out in detail and with complete mapping of regulator, operator, and structural gene, it does appear that there are nonetheless, genes for the control of other genes. We may therefore logically use in the study of higher creatures the knowledge of affairs in *E. coli*

There is one way in which the facts about *coli* may be at once extended to help in our thinking about higher creatures. We have seen how a single regulator gene can regulate many genes, the principle being merely that all of these genes have the same operator. We may think in principle of regulator genes being similarly grouped into larger categories as outlined in Fig. 2. According to this concept (Waddington, 1962) a single regulator, R_A acts upon the operators O_1, O_2 etc. of a whole series of regulators R_1, R_2 etc. Each regulator subgroup, as R_1, in turn acts upon a series of structural genes S_a, S_b, etc. through their operators O_a, O_b, etc. The regulation of each category S_a, S_b, by its regulator R_1 is modulated by a specific small molecular effector. The regulation of the whole class of regulators R_1, R_2, etc. by super-regulator R_A is in turn modulated by its own small molecular effector. It will be immediately clear how such categorization of control will be useful in our thinking about control of differentiation. The genome says to itself for example, 'Genes of category R_1, R_2, etc. may function under conditions 1, 2, etc. but only if the further condition is met that regulator A finds itself in condition A'. This is exactly the kind of model needed for understanding our example of flowering. The genes for flowering will be under control of regulators, all of which respond to the flowering signal, but only when the super-regulator, R_A, which regulates them, finds itself in a bud and not say in a root.

So much for the genetics of gene regulation. The studies on *E. coli* have given us a model of extraordinary power. The model is, however, merely a wiring diagram; we have not as

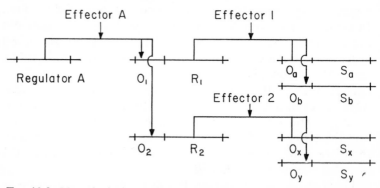

Fig. 10.2. Hypothetical model of arrangements suitable for categorization of control units. The ability of effectors 1, 2, etc. to modulate the repression exerted by regulator genes R_1, R_2, etc. is in turn modulated by the presence or absence of effector A.

yet appended any hardware to it. No molecular names have been given to the entities other than to the genes themselves which we must presume to consist of DNA. What is the chemical nature of the repressor, the substance made by the regulator gene which in turn acts upon the operator genes? It is argued by Monod *et al.* (1963) that the repressor substance must be a protein since it must combine specifically with a small molecule such as a β-galactoside or arginine or tryptophane and change its conformation as a result. Only proteins are known to be able to combine with specific small molecules and, as a result, to change dramatically in properties. Such allosteric proteins generally contain two to several peptide chains. In the most dramatic case the union of monomers to form the protein polymer depends completely upon the presence or absence of a particular substrate. An enzyme which we have studied, an

ATPase, is, in the absence of ATP, a polymer consisting of several subunits fastened together. In the presence of ATP each subunit binds some of it. The polymer falls to pieces into monomers as a result (Ts'o *et al.*, 1957). This is exactly the kind of change in conformation which we would like to imagine in our repressor protein. It must in the one state do something quite different from what it does in the second.

Let us look more generally at the tasks of the repressor substance. Firstly it must bind to some particular species of small molecule and so be altered in configuration. Secondly it must recognize particular operators; that is, it must be able to detect information written in DNA language. This implies that the repressor must possess the power of recognition based upon base complementarity and base pairing. It would seem sensible that the regulator substance should contain single-stranded nucleic acid, for example, RNA.

It should be noted that the number of effector substances in an organism may very well be considerably smaller than the number of operator sites. This is because each operator must have its own specific repressor whereas a number of repressors may share one effector substance in common.

We have then a control problem which requires both the subtle dexterity of protein and the subtle literacy of nucleic acid. We at once ask, could the regulator then be a complex of protein and RNA—protein to sense the effector molecule, RNA to sense the operator? Once we consider the problem in this light it becomes clear that this double-headed monster hypothesis is supported by some evidence.

The hypothesis is, firstly, merely an extension of an established biological stratagem. We know how nature has solved the analogous problem of the recognition of base sequences by amino acids. Nature uses an adapter of transfer RNA, a molecule which states at one end its name in nucleic acid language and at the other end its name in enzyme language. To be sure, we do not know how this is achieved, but the result is that a given transfer RNA can complex with but one amino acid activating enzyme and can deliver its amino acid only to one position, that specified by a particular nucleic acid code word. We could now extend the principle of adapters to hypotheses concerning the recognition of operator genes.

It might be imagined, that the regulator gene is transcribed into an RNA which at one end will base pair with a particular operator and at the other end will combine with a protein of some particular nature and itself able to combine with a particular effector. The number of kinds of proteins need not be as large as the number of operators but merely as large as the number of effectors, a few tens or hundreds. According to this view the repressor protein will not be produced by transcription of the messenger RNA of the particular regulator gene. There will rather be in the genome genes which synthesize the necessary repressor proteins. We would expect to find mutations of such genes, mutations which would then result in lack of regulation by whole classes of regulators and the operators for which they are responsible. Such mutations have not yet been recognized.

Let us approach the subject of the material nature of the repressor by first considering micro-organisms. That the repressor for the gene in *E. coli* responsible for the synthesis of alkaline phosphatase (a gene repressed in the presence of high concentrations of orthophosphate) is a protein has been shown by Gallant and Stapleton (1964). The repressor decays during derepression and must be synthesized anew under conditions requiring repression. This synthesis is inhibited by inhibitors of protein synthesis such as chloramphenicol. It is selectively inhibited also by canavanine, an analogue and competitive inhibitor of arginine. Synthesis of the repressor is in fact so much more sensitive to the presence of canavanine than is enzyme synthesis in general, that the gene for alkaline phosphatase can be derepressed simply by growing the cells with canavanine. Gallant and Stapleton suggest that the repressor may be a protein rich in arginine, i.e. a basic protein. Other equally forceful experiments suggest however that the repressors of *E. coli* consist of RNA. Thus both Sypherd and DeMoss (1963) and Bowne and Rogers (1963) have found that in the presence of chloramphenicol, which inhibits protein synthesis and causes accumulation of RNA, repressor also accumulates. Thus, depending on the bacterial strain, growth conditions, and gene studied, the repressors of *E. coli* appear to be either RNA or protein. What is one to conclude? A possible conclusion might be that the repressor consists of both protein and RNA. This has been in fact suggested by Sypherd and Strauss (1963).

Clearly though it is but a matter of time before a bacterial repressor will be isolated and its composition determined.

We have already seen that in the case of chromatin histones are one large class of repressor substances. Is the histone merely the protein part of a repressor which includes also an RNA molecule as an operator decoder? This question is not yet answerable. It is true that chromatin as isolated does contain some RNA (Chapter 2) and that a part of this RNA accompanies the nucleohistone component during the shearing of chromatin. Since the nucleohistone component does not support RNA synthesis, this RNA cannot be messenger RNA that was being synthesized when the biologist ground up the cell. The chromosomal RNA possesses two other interesting characteristics. When chromatin is fractionated by 4 M cesium chloride, in which nucleic acid sinks while protein floats, the chromosomal RNA accompanies the histone. Since it is not separable from the histone by usual RNA solvents, the RNA may be bound in some way to the histone. In addition, the chromosomal RNA is bound in a form in which it is almost totally resistant to attack by RNAase (Bonner *et al.*, 1962). From this bound form, RNA may be released by DNAase which shows that the RNA is in some way stabilized in chromatin by DNA. The chromosomal RNA may be released too by treatment at a temperature, 60°, below the melting temperature of the chromosomal DNA. Such treatment is attended by some derepression.

Quite evidently the chromosomal RNA is worthy of study. Equally evidently its role is at present unassessable. Let us therefore leave this matter and go on to a further one. To forward our understanding of the control of gene activity it will be desirable to focus upon a particular gene together with its operator, its regulator and specific repressor, and its effector, the small molecule which combines with the repressor to alter its conformation and causes derepression, or alternatively repression. The gene for globulin synthesis in pea cotyledons is, thus far at least, inappropriate to our purpose since we do not know of its effector. We do not know the signal in response to which globulin making is turned off and on. We have chosen, rather, to try to find an effector which acts upon a repressor which acts upon a gene which produces a messenger RNA which codes for an enzyme which can be identified. This is, as one can imagine, a long task.

What are the effector substances which by their interplay on the genome bring about development? In micro-organisms the effectors are, of course, cellular metabolites, and repression merely serves the end of making only needed enzymes. Higher creatures possess this sort of regulation too but this is not the kind of regulation which determines the course of development. One great class of the effectors of development are the hormones. Let us examine two examples.

We choose our first example on the basis of its drama. Let us pose ourselves the problem of finding cells in which the genome is totally repressed, and then try to find out what substances will derepress some portion of the genome. Where can cells be found in which the genome is totally repressed? An immediate thought is the erythrocytes of birds which are almost completely repressed. Very well, but again no derepressing agent has been found. Probably nature does not supply one. Erythrocytes are a dead end; they are not to become derepressed but are to die. There is, however, known to the plant biologist a phenomenon called dormancy. A cell tissue or organ which is dormant is alive and respires, but does not grow, even with all the conditions ordinarily favourable for growth. We know, for example, dormancy of buds. A dormant bud stays alive but it does not grow until some external signal appropriate to the particular kind of bud, a signal from the outside world, appears and says 'Now it's time to stop being dormant'. The buds of freshly-harvested potato tubers, for example, are dormant and begin to grow only after several months. We can at any time supply to the dormant potato buds the hormone, gibberellic acid, or alternatively, ethylene chlorhydrin which mimics the action of gibberellic acid in this function and a few days later the buds start to grow. The basis of this dormancy has been established by Tuan (1964). The genome of the dormant potato bud is essentially completely repressed. Not only does it in life lack the ability to synthesize RNA at any appreciable rate, but the chromatin isolated from dormant potato buds lacks the ability to support DNA-dependent RNA synthesis in the presence of added polymerase. Non-dormant buds whether non-dormant by virtue of passage of time or by virtue of application of hormone, both synthesize RNA *in vivo* and yield chromatin capable of supporting DNA-dependent RNA synthesis. Tuan has followed in detail the kinetics of the derepression elicited by ethylene chlorohydrin. The initial function of the hormone is, as one would expect of an effector, one of derepression. RNA synthesis starts. From the switching on of such synthesis flow the other attributes of growth, protein synthesis, DNA replication, cell division, and so on. It has been mentioned above that chromatin is derepressed to some extent by being heated below the melting temperature of DNA. It is interesting that it has been known for many years that the dormancy of buds may be broken by soaking them in hot water, 50–60°, for a few minutes. This is another straw in the wind.

This example shows the cleverness with which chromosomal control is used in nature but does not provide a single gene with which to work since many genes are involved. Our next case however, will do just this.

The barley plant contains a gene for making α-amylase in the endosperm of the seed. This gene is fully repressed before seed germination. As the seed germinates, the growing embryo sends to the endosperm a signal which derepresses the α-amylase gene. Alpha amylase pours out of the endosperm. This is the basis of malting and hence of beer. The effector substance for the derepression of the α-amylase gene was discovered by Paleg (1960) to be gibberellic acid. If we cut off the embryo from the seed and soak the remaining embryoectomized endosperm, no α-amylase is made. If we give to such barley endosperm a small amount of gibberellic acid, α-amylase pours out. The phenomenon has been studied in detail by Varner and Ram Chandra (1964). The production of the enzyme is supported by DNA-dependent RNA synthesis since it is inhibited by actinomycin-D. The first detectable metabolic effect of the effector, gibberellic acid, is enhanced RNA synthesis. In this system we have the desired qualities: a repressed gene, its effector substance and an easily identifiable enzyme product of the structural gene. But most importantly, we have learned that a hormone can act as an effector substance, a small molecular agent of derepression. Is it possible that other hormones accomplish their work by acting as effector substances? With this question the following chapter will be concerned.

SELECTED REFERENCES

Genetics of genetic control mechanisms

JACOB, F. and MONOD, J., *J. mol. Biol.* **3,** 318 (1961).

HOROWITZ, N., FLING, M. MACLEOD, H. and WATANABE, Y., *Cold Spring Har. Symp. quant. Biol.* **26,** 233 (1961).

HALVORSON, H. O., *Idem* **26,** 231 (1961).

PONTECORVO, G., *Proc. Roy. Soc.* B **158,** 1 (1963).

RHODES, M., *Cold Spring Harb. Symp. quant. Biol.* **9,** 138 (1941).

McCLINTOCK, B., *Idem* **21,** 197 (1956).

McCLINTOCK, B., *Brookhaven Symp. Biol.* **8,** 58 (1956).

LEWIS, D., *Proc. Roy. Soc.* B **151,** 468 (1960).

WADDINGTON, C. H., *New Patterns in Genetics and Development.* Columbia University Press (1962).

Molecular biology of microbial control mechanisms

KEPES, A., *Biochim. biophys. Acta* **76,** 293 (1963).

HARTWELL, L. and MAGASANIK, B., *J. mol. Biol.* **7,** 401 (1963).

MONOD, J., CHANGEUX, J. P. and JACOB, F., *J. mol. Biol.* **6,** 306 (1963).

BOWNE, S. W. and ROGERS, P., *Biochim. biophys. Acta* **76,** 600 (1963).

SYPHERD, P. and DEMOSS, J. A., *Idem* **76,** 589 (1963).

SYPHERD, P. and STRAUSS, N., *Proc. Nat. Acad. Sci., Wash.* **50,** 1059 (1963).

GAREN, A. and GAREN, S., *J. mol. Biol.* **7,** 13 (1963).

GALLANT, J. and STAPLETON, R., *J. mol. Biol.* **8,** 431 (1964).

GALLANT, J. and STAPLETON, R., *J. mol. Biol.* **8,** 442 (1964).

Allosteric proteins

MONOD, J. CHANGEAUX, J. P. and JACOB, F., *J. mol. Biol.* **6,** 306 (1963).

TS'O, P. O. P., EGGMAN, L. and VINOGRAD, J., *Archs. Biochem. Biophys.* **66,** 64 (1957).

Molecular biology of chromosomal control mechanisms

BONNER, J., HUANG, RU-CHIH and MAHESHWARI, N., *Proc. Nat. Acad. Sci., Wash.* **47,** 1548 (1962).

VARNER, J. and RAM CHANDRA, G., *Proc. Nat. Acad. Sci., Wash.* **52,** 100 (1964).

TUAN, D. and BONNER, J., *Plant Physiol.* **39,** 768 (1964).

PALEG, L. *Plant Physiol.* **35,** 293 (1960).

11

HORMONES AS EFFECTOR SUBSTANCES

IT IS commonplace amongst biologists to think of hormones, or some of them, as agents which have to do with differentiation, agents which guide or at least evoke pathways of development. The induction of secondary sex characters in animals as a result of estrogen or androgen application is a type example of one kind of control of differentiation. In fact the ideal experiment in differentiation would be one in which we mix together various kinds of molecules, squirt in some androgen, and thus make a cock's comb grow magically out of the test tube. Or to take another example—we mix various other kinds of molecules, add some flower hormone, and as a result, a beautiful flower grows out of the test tube. The principal bar to the study of differentiation by way of hormones has been, however, that until recent times, not the slightest progress had been made in finding out how hormones do their work.

This unhappy state of affairs has now altered dramatically. The world is being drenched by a flood of revelation concerning the nature of hormone action. Indeed it seems possible that more biologists will be converted to molecular biology through this insight into the nature of hormone action than through any of the other insights into the nature of life which have been offered by molecular biology. Let us therefore scan the hormone world and find what our colleagues have discovered. In summary, they have discovered that hormones work by derepressing genes previously repressed, thus causing synthesis of messenger RNA molecules not made in the absence of the hormone and thus in turn causing the synthesis by the tissue or organ of new kinds or of increased amounts of enzyme molecules.

It has been known for many years that when rats or other mammals are treated with cortisone, the levels of several enzymes in the liver rise dramatically (Knox et al., 1956). Among these is glutamic-tyrosine transaminase. Kenney (1962) has isolated the purified enzyme from the liver of rats given, not only cortisone, but also labelled amino acid. The isolated enzyme was labelled, showing that it had been synthesized from its constituent amino acids as a result of cortisone application. Cortisone then causes the *de novo* synthesis of a particular enzyme, but, how does it do so? First of all the activities of several liver enzymes are increased as a result of cortisone treatment. These include, in addition to the glutamic-tyrosine transaminase, glutamic-alanine transaminase, trypto-phan pyrolase, glucose-6-phosphatase, fructose diphosphate phosphatase, and aldolase among others. The activities of these enzymes are increased in the liver by two-to tenfold in response to cortisone treatment. The appearance of the enzymatic activity under the influence of cortisone is suppressed if the creature is simultaneously given the magic substance puromycin. Puromycin, an antibiotic, is an amino acylnucleo-side which by mimicing the action of a proper amino acyl

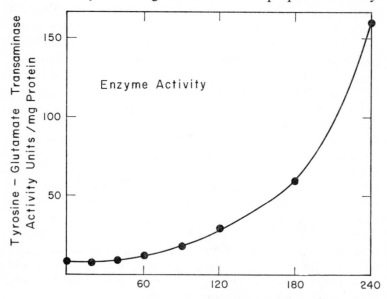

FIG. 11.1a. Kinetics of cortisone-induced appearance of tyrosine-glutamate transaminase and of increase in rate of nuclear RNA synthesis.
(a) Time course of appearance of the enzyme as a function of time after injection of cortisone.

transfer RNA breaks the growth of the peptide chain upon the ribosome. Puromycin is hence an elegantly specific inhibitor of protein synthesis (Yarmolinsky *et al.*, 1959). We may conclude that all the cortisone-induced enzymatic activities of liver are owing to synthesis of new protein.

The kinetics of appearance of the increased enzymatic activity are simple. It commences about 2 hours after injection of cortisone into the rat and continues indefatigably, reaching a plateau after some 5 days (Segal and Kim, 1963). The time for one half of maximal effect to be achieved is 1·2 days. After cessation of daily cortisone injections, the increased enzymatic activities die away with a half-life of 3·5 days—about that of turnover of liver protein as a whole.

More interestingly, the treatment of rats with cortisone

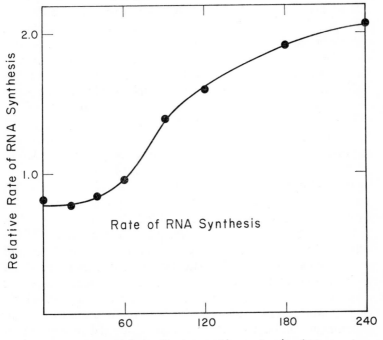

FIG. 11.1b. (b) Rate of nuclear RNA synthesis as a function of time after injection of cortisone. Each point represents amount of P³² incorporated into RNA in 20 minutes (after Kenney and Kull, 1963).

alters the rate at which RNA is synthesized in the nuclei of their liver cells (Kenney and Kull, 1963). Within 1 hour after cortisone injection, rate of RNA synthesis in the nuclei of the liver cells begins to increase, reaching a plateau of about twice the control rate after some 4 hours. This early increase in rate of nuclear RNA synthesis is followed after some hours by an increasing rate of appearance of new RNA in the cytoplasm of the liver cells. Such RNA, made by liver in response to cortisone injection, has been subjected to sucrose density gradient centrifugation (Kenney and Kull, 1963) and been found to possess a broad range of sedimentation coefficients including those characteristic of messenger RNA. And most importantly, the synthesis of RNA elicited by cortisone is essential to the appearance of the enzymatic activities which are also elicited by cortisone. That this is so is shown by the use of actinomycin-D. This substance base pairs specifically with guanine and in fact only with guanine as it is found in double helical DNA. The conformation of this complex is such that the actinomycin-D sticks out from the DNA helix axis like a banderilla in a bull and by so doing prevents the RNA polymerase molecule from using the DNA chain as a template. Actinomycin-D specifically and completely blocks DNA-dependent RNA synthesis in the test tube (Hurwitz *et al.*, 1962), in bacteria (Levinthal *et al.*, 1962), and in animal cells (Franklin, 1963) and this in concentrations of the order of 10^{-6} M. Not only does it abolish cortisone-induced increases in rate of liver RNA synthesis (as well as of all liver RNA synthesis) but it also abolishes the induction of liver enzymes by cortisone (Greengard and Acs, 1962; Weberg *et al.*, 1963).

We have then a rather complete physiological picture of the action of cortisone on liver. Cortisone goes to the liver and somehow causes more of particular kinds of messenger RNA to be made, that is, it derepresses. This messenger RNA is used for the conduct of enzyme synthesis. In a particularly elegant work, Segal and Kim (1963) have done, on the rat, an experiment similar to that previously discussed in connection with chromatin-supported pea-seed globulin synthesis. Their experiment consists in finding the preportion which a particular cortisone-induced enzyme constitutes of the total protein synthesized by liver in the cortisoneless and in the cortisone-treated state. Rats were injected or not injected with cortisone and one day later supplied with C^{14}-leucine. After an incubation period of one hour, rats were killed, their livers removed, the soluble proteins of the liver prepared, and glutamic-alanine transaminase isolated by a combination of enzymological and immunochemical methods. The data of Table 1 show that the glutamic alanine transaminase constitutes but 0·24 per cent of total soluble protein synthesized in liver of rats not treated with cortisone. Since Segal and Kim had no control on non-specific background contamination of the enzyme-antibody

TABLE 1

Glutamic-alanine transaminase synthesis by rat liver as a function of prior treatment (24 *hours before*) *with a single injection of cortisone. Enzyme isolated by immunochemical procedures. Incubation* (*in vivo*) *for 1 hr*

	Leucine incorporated into		
Treatment of animals	Total soluble protein (cpm)	glu-ala transaminase (cpm)	glu-ala transaminase as % of total protein
Control (no cortisone)	19×10^3	45	0·24
Cortisone	24×10^3	298	1·3

After Segal and Kim, 1963.

precipitate, the true figure may be smaller. We have seen with the immunochemical isolation of pea-seed globulin that there is an irreducible background contamination of 0·1 to 0·2 per cent. The enzyme constitutes however 1·3 per cent of total soluble protein formed by livers of cortisone-treated rats.

We have then every link in the chain of evidence:—cortisone derepresses particular genes in the genome of liver cells. Cortisone is then an effector substance, but how does cortisone effect its effect? We do not know; the physiologists have not gotten to this yet. Obviously, however, we are now in a position to discover whether cortisone can derepress liver chromatin in the test tube. If the answer to this question is 'yes' we will then possess a system for the elucidation on the molecular level of how effector effects.

Cortisone is, of course, formed in the adrenal cortex. There it is manufactured by appropriate enzymes. The enzymatic manufacture of cortisone in the adrenal cortex is under the control of a pituitary hormone, adrenocorticotrophic hormone, ACTH. ACTH goes from the pituitary to the adrenal cortex and says 'Please make me some cortisone'. How does ACTH do its work? We would imagine that the ACTH again is an agent of derepression; that it, in the cells of the adrenal cortex, derepresses those genes which make the messenger RNA for making the enzymes for making cortisone. We do not know this as fact, but a straw in the wind is provided by Farese and Reddy (1963). They supplied ACTH twice daily to rats and followed the biochemistry of the adrenal cortex. Both protein and RNA content of the tissue increased within one day after the initiation of hormone treatment. RNA rose to three times the initial level and protein by fourfold. It has not yet been shown that particular enzymes in the adrenal cortex increase in response to ACTH treatment or that such increase, if it takes place is actinomycin-D inhibitable. This remains for the future. For the present, we may guess that ACTH is probably also an agent of derepression.

Of all of the hormones, those related to sex have been the most favoured for physiological study. This fact no doubt possesses some significance in its own right. We are now coming to know something of how the estrogens do their work. Firstly, the injection of an estrogen into an immature or ovariectomized rat brings about characteristic histological changes—alterations in the uterine wall and proliferation of the vaginal mucosa. Within four hours after estrogen treatment the rate of protein synthesis in the cells of the uterine wall rises (Mueller et al., 1961; Hamilton et al., 1963). This increased rate of protein synthesis is preceded by a burst of RNA synthesis (Ui et al., 1963; Noteboom and Gorski, 1963; Wilson, 1963). This is a large effect—rate of RNA synthesis is increased by over tenfold. Such estrogen-induced RNA synthesis, as well as the induced protein synthesis, is abolished by topical treatment with actinomycin-D; both are DNA dependent. And finally, an elegant start in the further analysis of how estrogens work has been made by Wilson (1963). She injects estrogen into previously ovariectomized rats, removes their uteri, and incubates slices of such uteri in media containing labelled RNA precursors. The slices respond by major increases in rate of RNA synthesis just as does the tissue in the intact animal. RNA from such estrogen-treated uteri was supplied to liver ribosomes and shown to exhibit messenger RNA activity. The RNA of uteri of animals not given estrogen treatment did not possess as much messenger RNA activity. So again, we see that the hormone does its work by eliciting the synthesis of messenger RNA which then elicits protein synthesis, the synthesis of new enzymes whose identity we do not know, which then in turn elicit changes in cell shape, size, and function. Estrogens also would appear to be effector substances. Whether estrogens act directly on the chromosome has not yet been tested. Potentially, however, we do have in the uterus a possible system for the study of the way in which effector substances effect.

Similar relations hold for the androgens and their target organs—testes, prostate gland, and seminal vesicles among others. Butenandt and his colleagues (1960) first found that injection of testosterone into immature or castrated rats causes a fourfold increase in rate of protein synthesis in the target organs. Liao and Williams-Ashman (1962) have now shown that testosterone treatment increases rate of RNA synthesis in the prostate and that this RNA confers upon the ribosomes of the prostate the ability to synthesize protein—that is, it is messenger RNA.

The world is abrim with oddments concerning testosterone— how its effects are actinomycin-D inhibitable, how its ability to elicit protein synthesis is inhibited by puromycin, etc. The effects of the hormone are then due to new protein synthesis mediated by DNA-dependent RNA synthesis. The very synthesis in the testes of the androgenic hormone is itself hormonally controlled by the gonadotrophic pituitary hormone, and in this function, the action of the gonadotrophic hormone is inhibited by actinomycin-D (Talwar and Segal, 1963).

It is, in fact, becoming increasingly difficult to find a hormone which does not appear to work by derepressing the genetic material. Take, for example, insulin. It has been known since 1952 (Sinex et al., 1952) that the addition of insulin to isolated rat diaphragm muscle increases the rate of protein synthesis. It was subsequently discovered that increased rate of RNA synthesis is associated with the action of insulin on this tissue and now Wool and Munro (1963) have shown that the incremental RNA whose synthesis is elicited by insulin, has the sedimentation characteristics of messenger RNA. This is however as far as molecular biology has gone in the investigation of insulin action. The implications are clear, however, that insulin must do its job of affecting carbohydrate metabolism by controlling the synthesis of enzymes which have to do with the pathways of carbohydrate utilization. We add insulin to our fold.

The action of thyroxine too, is inhibited by actinomycin-D and puromycin as shown by Tata (1963) who followed the influence of thyroxine in thyroidectomized whole rats. We use the whole living rat since the active form of the hormone is unknown and in addition it affects cells and tissues generally. The effect of thyroxine on rats was studied by measurements of their metabolic rate and of their rate of growth. The administration of thyroxine increased rate of metabolism by some 35 per cent while actinomycin-D by itself decreased it by some 14 per cent. The administration of thyroxine and actinomycin-D together, caused the metabolic rate to remain at the actinomycin-D level. The effect of thyroxine in increasing metabolic rate is thus totally suppressed by actinomycin-D. The implication is clear. Similar findings were made on the effect of thyroxine in increasing rate of growth of thyroidectomized rats. In both cases too, the ultimate expression of the influence of the hormone must be through protein synthesis, since the effects of thyroxine both on metabolic rate and growth are not only inhibited by actinomycin-D but also by puromycin.

Let us turn to the growth hormone. This protein hormone is produced by the pituitary and is required for mammalian growth. The effects of growth hormone may be conveniently studied in hyphophysectomized rats which cannot make it. The administration to such hyphophysectomized rats of growth hormone results increases rate of RNA production by the liver as shown by Talwar and his colleagues (1962) and by Korner (1963). The increased rate of RNA synthesis is, of course, actinomycin-D suppressible. In addition, growth hormone causes increased rate of liver protein synthesis and this is mirrored in increased rate of incorporation of labelled amino acids into protein by isolated liver ribosomes. The ribosomes of liver of growth hormone-treated rats possess a higher capability for protein synthesis *in vitro* than do the ribosomes from liver of growth hormone non-treated rats. These observations are perhaps but straws in the wind. They are not yet complete and rigorous. But again they suggest that the action of the growth hormone, like that of the other pituitary hormones —adrenocorticotrophic hormone and gonadotrophic hormone —has to do with the production of messenger RNA in the target organs. The evidence that is lacking in all of these cases is first identification of particular enzymes which are produced in increased amounts as the result of hormone treatment and secondly, evidence concerning whether or not the hormone does its work by interacting directly with the chromosomal material of the cell.

Even the tumorologists are swept along in the excitement of the day. It has been known since the olden days of chimney-sweeps in Britain, that polycyclic aromatic hydrocarbons are carcinogenic and that in rats, they produce tumours, in, for example, the liver. Gelboin and Blackburn (1963) have found that administration of a typical carcinogenic hydrocarbon, 3-methylcholanthrene, to rats causes the appearance in the liver of particular enzymes. Other kinds of enzymes remained unchanged in concentration. One enzyme, benzpyrene hydroxylase, typical of those whose amount increases as a result of methylcholanthrene administration, has been picked for further study. This enzyme increases in activity approximately twenty-fold after administration of the carcinogen. The increase does not occur in the presence of actinomycin-D. Ribosomes from livers of methylcholanthrene-treated rats are also more active in protein synthesis than those from normal rats implying that they are better supplied with messenger RNA. The carcinogen modifies in some way the normal pattern of genetic repression.

We move now to still a different manifestation of hormonal activity—that of the insect moulting hormone—ecdysone. The moulting, between one insect larval instar and the next, is controlled by ecdysone—a sterol hormone, cholesterene—produced in the prothoracic glands of the larva (Karlson, 1963). These in turn, by their periodic output of ecdysone, control the periodicity of moulting. The target organs of the hormone are

9

many, perhaps all the cells of the larva. Fortunately ecdysone regulates the moulting of Dipteran larvae for these possess a useful cytological peculiarity—namely the giant, polytene chromosomes of their salivary glands. Such giant chromosomes consist of some 1000–4000 chromatids, or single chromosomal fibres which lie side by side in huge bundles—all strands in register. Aided by these circumstances, the cytologist can recognize a regular, and for each chromosome, characteristic, succession of bands and interbands. We know, from the vast mass of Drosophilogical genetics and cytology, that each band is a gene or a small group of genes and that in the polytene chromosome, we see before our eyes, the linear array of the genetic units.

During the course of larval development, particular and characteristic chromosomal bands become swollen while others remain unswollen. Such swelling, called puffing, is as shown by Pelling (1959) associated with RNA synthesis. That the RNA synthesis associated with puffing is DNA-dependent is shown by its inhibition by actinomycin-D. We imagine then, that puffs are derepressed genes—non-puffs repressed genes. The act of moulting is itself associated with the puffing in characteristic time-sequence of specific chromosomal bands. This same sequence may be elicited by the injection of ecdysone into young larvae—that is larvae immediately after a moult and therefore before the normal time for a further moult. Thus within 15 minutes of the injection of ecdysone, band No. 1-18-C of the *Chironomus* larva puffs and this is followed 15 minutes later by the puffing of band 4-2-B (Clever and Karlson, 1960; Clever, 1961). Puffing under the influence of applied ecdysone is also inhibited by actinomycin-D and is therefore DNA-dependent. The RNA products of different puffs have different base compositions as one would expect of the products of different genes (Edstrom and Beermann, 1962).

The insect pupal moult involves the production by the epidermis of the larva of the hardening agent, N-acetyl dihydroxyphenylethylamine. The latter in turn is formed from dihydroxyphenylalanine (DOPA) by the enzyme DOPA decarboxylase. Formation of this enzyme in the larval epidermis is induced by treatment with ecdysone. Sekaris and Lang (1964) have treated Calliphora larvae with ecdysone, isolated the nuclei of the epidermal cells and from these nuclei have isolated nuclear RNA. This RNA, which includes messenger RNA, was then supplied to a messenger RNA-dependent liver ribosomal protein synthesizing system. They have shown that messenger RNA from ecdysone treated larvae causes the production by the protein synthesizing system of detectable amounts of DOPA decarboxylase. Messenger RNA from control larvae not treated with ecdysone does not elicit the synthesis of DOPA decarboxylase by the ribosomal system. This elegant experiment shows directly that the effect of hormone treatment is to cause production of particular new species of messenger RNA.

The ecdysone-salivary gland chromosome system is then a suitable one for the study, on the level of cytology, of how effector-induced derepression works. One of the first questions to be answered is whether ecdysone binds to the chromosome or binds to something which then floats off the chromosome. Further questions concern chromosome structure. In the unpuffed band the genetic nucleohistone is condensed into a compact, much coiled form. Puffing involves loosening of the structure. What are the actual molecular structures in the two states, puffed and unpuffed, and how is this change in structure related to derepression? Is the tightly coiled structure the basic cause of repression as implied by Hsu (1962) or is its unfolding merely a necessary prerequisite to subsequent specific derepression as implied by the work of Fujita and Takamoto (1963)?

That hormones act as effectors of genetic derepression is also clear for the case of plants. We have discussed earlier the example of dormant tissues and organs—the cells of which lie inert even in the presence of all the conditions favourable to growth—good temperature, presence of water and nutrients, etc. We saw how the dormancy of buds, which are in nature called to growth by one or another environmental factor—long days, low temperature, or mere passage of time—can be broken by the administration of the plant hormone, gibberellic acid, or of a substance which mimics its action, ethylene chlorohydrin. We saw that in dormant buds the genetic material is almost totally repressed and that the dormancy-breaking agent derepreses genetic material previously repressed. Not only do the treated buds begin to grow and to conduct DNA-dependent RNA synthesis but their isolated chromatin supports

DNA-dependent RNA synthesis in the presence of added exogenous RNA polymerase which the chromatin of dormant buds cannot. Very good. In this dramatic instance, we arouse the genome from a state of almost total repression to one of extensive derepression by administering a hormal effector substance.

But the further analysis of dormancy is difficult. We do not know any of the individual enzymes whose messenger RNA's are produced under the influence of the derepressing effector substance and the logistics of dormant buds are unfavourable. It would take seven maids with seven pairs of hands more than half a year to dissect out enough dormant buds to supply enough chromatin to investigate the interaction of chromatin and effector substance. We turn, therefore, to still another system involving a plant hormone. This is the hormonally induced synthesis of α-amylase by the aleurone layer of barley seeds discussed in the previous chapter. In response to gibberellic acid application, α-amylase is synthesized out by the aleurone layer in massive quantities. Let us consider this system as its characteristics have been revealed by Varner and Ram Chandra (1964). In the first place, application of gibberellic acid to the barley aleurone almost immediately doubles its rate of RNA synthesis. The increased RNA synthesis is actinomycin-D inhibitable and is therefore DNA-dependent. The RNA synthesized by the aleurone layer under the auspices of gibberellic acid exhibits a sedimentation pattern characteristic of messenger RNA—it is not ribosomal RNA. The synthesis of α-amylase in response to gibberellic acid-treated aleurone layers is suppressed by actinomycin-D and is therefore supported by newly formed messenger RNA. And finally, α-amylase has been isolated in pure form from gibberellic acid-treated aleurone layers supplied simultaneously with labelled amino acid. The α-amylase thus isolated is itself labelled. It is therefore made *de novo*. The α-amylase synthesized under the direction of added gibberellic acid has been subjected to tryptic hydrolysis and the resulting peptides separated by chromatography. In this way the gibberellic acid-induced α-amylase has been shown to be identical with that produced by normally germinating barley seeds.

The aleurone layer of barley seeds synthesizes both RNA and protein sluggishly. We may therefore guess that the genome of aleurone cells is largely repressed. As a result of gibberellic acid application, both rate of RNA synthesis and of protein synthesis are approximately doubled and it is found that α-amylase constitutes almost one half of the total protein formed. Clearly then, the hormone, gibberellic acid, acts as should an effector substance. It causes a gene previously repressed to become derepressed. It causes the target organ to produce a specific enzyme in vastly increased proportion of total protein synthesized. Clearly, all that remains to be done is to discover the mode of interaction of hormone with the chromatin itself. To do this, we must have pure chromatin and this has proven to be a difficult task. The aleurone cells contain large and dense protein bodies—the aleurone grains, which are difficult to separate from chromatin.

The plant growth hormone, indoleacetic acid, like gibberellic acid, controls cell elongation and causes actinomycin D-inhibitable increases in RNA and protein synthesis (Key, 1964). Although the enzymology by which the hormone exerts its control of growth is unknown, we do know of one enzyme, probably not related directly to the growth process, whose abundance in plant tissue is controlled by indoleacetic acid. This enzyme is indoleacetyl aspartate synthetase which catalyzes the synthesis of indoleacetyl aspartate as well as of related components as benzoyl aspartate. This enzyme is normally present in pea stems in small amounts. When sections of pea stems are soaked in a solution containing a low concentration (10^{-5} M) of indoleacetic acid, the enzyme immediately begins to appear in the tissue in increased amounts. Venis (1964) has shown that the new production of the enzyme is suppressed by puromycin and by actinomycin D. The hormonally induced enzyme synthesis is therefore supported by DNA-dependent RNA synthesis. In this case also the hormone clearly acts as an effector of derepression.

To give us a feeling for the kind of phenomena which we are likely to meet in our further studies of the hormones as effector substances, let us consider one last example—that of the transformation by the flowering hormone of the vegetative bud of the plant into a flower bud. The flowering hormone, which appears to be a steroid (Bonner et al., 1963) is synthesized in the leaves in response to an external signal. The hormone travels to the bud where, in the course of a few hours, it does something which commits the bud to turning into a flower, a process which becomes visible microscopically some twenty-four hours later. What the hormone does to the bud may be described in terms of molecular biology as follows. The hormone says to its target cells 'Genes for making flowers, fruits, seeds, etc., you have been repressed up to now. Now, however, it is time for you to become derepressed; to make the messenger RNA's and hence the enzymes required for the production of the reproductive organs.' The histology of the arrival at the bud of the flowering hormone is dramatic, as has been demonstrated by Gifford and Tepper (1962). Shortly after receipt of the flowering message by the bud and before any visible signs of differentiation are apparent, a decrease in histone content of the cells of the bud takes place. This is followed by an increase in RNA concentration in the same cells, an increase due to an increase in rate of RNA synthesis. This is in turn attended by an increased rate of protein synthesis. We have in these facts a direct, albeit *in vivo*, demonstration of one way in which we may expect hormone and chromatin to interact. We know that we must expect derepression, and we see, in this instance, that derepression is accompanied by loss of histone from the chromatin. This accords with the views developed above of histones as the engines of repression.

The hormones then act in many cases, perhaps in all, as agents which control genetic activity. Through the study of hormone-genome interaction we may hope to gain understanding on the level of molecular biology of the nature of genetic switching devices.

SELECTED REFERENCES

Cortisone and related materials

KNOX, W. E., AUERBACH, U. H. and LIN, E. C., *Physiol. Rev.* **36**, 164 (1956).

KENNEY, F., *J. Biol. Chem.* **237**, 3495 (1962).

KENNEY, F. and KULL, F., *Proc. Nat. Acad. Sci., Wash.* **50**, 493 (1963).

SEGAL, H. and KIM, Y. S., *Proc. Nat. Acad. Sci., Wash.* **50**, 912 (1963).

GREENGARD, O. and ACS, G., *Biochim. biophys. Acta* **61,** 657 (1962).

WEBER, G., SINGHAL, R. L. and STAMM, N., *Science* **142,** 390 (1963).

ACTH

FARESE, R. V. and REDDY, W. J., *Biochim. biophys. Acta* **76,** 145 (1963).

Estrogens

MUELLER, G. C., GORSKI, J. and AIZAWA, Y., *Proc. Nat. Acad. Sci., Wash.* **47,** 169 (1961).

TALWAR, G. and SEGAL, S. J., *Proc. Nat. Acad. Sci., Wash.* **50,** 227 (1963).

WILSON, J., *Proc. Nat. Acad. Sci., Wash.* **50,** 93 (1963).

HAMILTON, T. H., *Proc. Nat. Acad. Sci., Wash.* **49,** 373 (1963).

UI, H. and MUELLER, G. C., *Proc. Nat. Acad. Sci., Wash.* **50,** 256 (1963).

NOTEBOOM, W. and GORSKI, J., *Proc. Nat. Acad. Sci., Wash.* **50,** 250 (1963).

SCHJEIDE, O. A. and WILKENS, M., *Nature, Lond.* **201,** 42 (1964).

Androgens

BUTENANDT, A., GUNTHER, H. and TURBE, F., *Z. Physiol. Chem.* **322,** 28 (1960).

RYAN, K. S., MEIGS, R., PETRO, Z. and MORRISON, G., *Science* **142,** 243 (1963).

SHAW, C. R. and KOEN, A. L., *Science* **140,** 70 (1963).

LIAO, S. and WILLIAMS-ASHMAN, H. G., *Proc. Nat. Acad. Sci., Wash.* **48,** 1956 (1962).

SILVERMAN, D. A., LIAO, S. and WILLIAMS-ASHMAN, H. G., *Nature, Lond.* **199,** 808 (1963).

Insulin

SINEX, F. M., MCMULLEN, J. and HASTINGS, A., *J. Biol. Chem.* **198,** 615 (1952).

WOOL, I. G. and MUNRO, A. J., *Proc. Nat. Acad. Sci., Wash.* **50,** 918 (1963).

Thyroxin

TATA, J. R., *Nature, Lond.* **197,** 1167 (1963).

Growth hormone

KORNER, A., *Biochem. biophys. Res. Commun.* **13,** 386 (1963).

TALWAR, G., PANDA, N. C., SARIN, G. S. and TOLANI, A. J. *Biochem. J.* **82,** 173 (1962).

Carcinogens

LOEB, L. and GELBOIN, H. V., *Nature, Lond.* **199,** 809 (1963).
GELBOIN, H. V. and BLACKBURN, N., *Biochim. biophys. Acta* **72,** 657 (1963).

Ecdyson-puffing

FUJITA, S. and TAKAMOTO, K., *Nature, Lond.* **200,** 494 (1963).
KARLSON, P., *Colloquium Ges. physiol. Chem.* **13,** 101 (1962).
CLEVER, U., *Devl Biol.* **6,** 73 (1963).
EDSTROM, J. and BEERMANN, J., *J. cell. Biol.* **14,** 371 (1962).
HSU, T. C., *Expl cell. Res.* **27,** 332 (1962).
SEKERIS, C. E. and LANG, N., *Life Sci.* **3,** 625 (1964).

Gibberellic acid

PALEG, L., *Plant Physiol.* **35,** 293 (1960).
EDELMAN, J. and HALL, M. A., *Nature, Lond.* **201,** 296 (1964).
VARNER, J. and RAM CHANDRA, G., *Proc. Nat. Acad. Sci., Wash.* **52,** 100 (1954).

Puromycin

YARMOLINSKY, M. B. and DE LA HABA, *Proc. Nat. Acad. Sci., Wash.* **45,** 1721 (1959).

Actinomycin

HURWITZ, J., FURTH, J., MALAMY, M. and ALEXANDER, M., *Proc. Nat. Acad. Sci., Wash.* **48,** 1222 (1962).
LEVINTHAL, C., KEYNAN, A. and HIGA, A., *Proc. Nat. Acad. Sci. Wash.* **48,** 1631 (1962).
FRANKLIN, R. M., *Biochim. biophys. Acta* **72,** 555 (1963).

Flowering hormone

BONNER, J., HEFTMANN, E. and ZEEVAART, J. A. D., *Plant. Physiol.* **38,** 81 (1963).
GIFFORD, E. M. JR. and TEPPER, H. B., *Am. J. Bot.* **49,** 706 (1962).
ZEEVAART, J. A. D., in *The Nucleohistones* (BONNER, J. and TS'O, P. O. P., Editors). Holden-Day, San Francisco (1964).

Auxin

CLICK, R. E. and HACKETT, D., *Fed. Proc.* **23,** 525 (1964).
KEY, J. L., *Plant Physiol.* **39,** 365 (1964).
KEY, J. L. and SHANNON, *Plant Physiol.* **39,** 360 (1964).
VENIS, M. A., *Nature, Lond.* **202,** 900 (1964).

12

SWITCHING NETWORKS FOR DEVELOPMENTAL PROCESSES

WE HAVE spoken of developmental processes as being guided by a properly programmed sequential and orderly repression and derepression of the individual structural units of the genome. That this is a correct way to speak of development as seen within the framework of molecular biology is evident from such considerations as the observed appearance of particular gene products only in particular cells at particular times during the developmental cycle. Let us consider further examples of such unleashing of previously leashed genetic information in relation to the induction in an organ of a new developmental pathway.

In an organ, as it pursues its original path of development, certain genes are derepressed while others are repressed. We now impress upon the organ circumstances which call for the initiation of an altered pathway of development. According to our picture of the nature of developmental processes, embarkation upon this altered pathway should require an alteration in the posture of the genome. Genes previously repressed must now be derepressed and perhaps, in addition, genes previously derepressed must be repressed. To this end we should apply the tool of RNA-DNA hybridization. By this means it would be possible to determine the extent to which the messenger RNA produced after induction differs from that produced by the same cells prior to induction; to determine whether embarkation upon a new developmental pathway depends upon the production of fresh species of messenger RNA. Unfortunately, however, no complete analysis, by hybridization studies, of the switching from one developmental pathway to another is yet available. We must therefore resort to other ways of discovering the nature of morphogenetic induction. One clear-cut demonstration that induction is accompanied by the transcription of genes previously repressed is that of Sekaris and Lang (1964) referred to in the previous chapter. Clearly, the induction of the moulting pathway involves the derepression of the gene for DOPA decarboxylase and the production of the messenger RNA for the assemblage of this enzyme.

In the analysis by Sekaris and Lang we sense that first hand, the genetic-switching machinery at work as the control settings responsible for the direction of one pathway, are reset to bring the morphogenetic progress onto a fresh course. Once set upon this new course, the further development will take place inexorably, step by step.

We must imagine that, as development continues, further switching will occur. Buds awakened by a single brief treatment with ethylene chlorohydrin develop into shoots with all the appropriate shoot parts. Vegetative buds induced by flowering hormone develop step by step into floral primordia, flowers, and fruits. Fresh genes must then be turned on and off at appropriate places and times to result in the appearance of the characteristic final form. This whole sequence of genetic switching is set in motion only as a result of the initial induction. Once induction has taken place, the further programming of development flows on automatically—as the acting out of a play whose text is already written in the genetic book. Induction is but the selection of the correct play.

To put it still another way, induction, embarkation upon a particular developmental pathway, consists in calling into action a pre-programmed subroutine, a subroutine containing in pre-programmed form the sequential steps and decisions required to bring about the required end result.

Let us further consider the concept of the life cycle as made up of a master programme constituted in turn of a set of subprogrammes or subroutines.

What will be required by way of subprogrammes to programme a creature through its life cycle? How might the master programme be divided into subprogrammes or subroutines? I choose a plant for consideration, and list in Table 1 the subroutines into which the whole genomal information might most obviously be partitioned. These include first, of course, a subroutine on cell life, how to make the subcellular organelles and enzymes needed for cellular growth, multiplication, and maintenance. Next we must have a subroutine on embryonic development together with one, perhaps a short one, on the general facts about life as a seed. Then, too, our plant requires subroutines containing information about the making of each of the kinds of vegetative organs—four in number for the plant. And finally, the plant must have a subroutine on reproductive development. These are sufficient to manage the life cycle of a modest plant. A more complex plant might require further information about catching insects or making spines etc. Even more, the additional information might be considered as merely an elaboration of one of the basic subroutines.

TABLE 1

Subroutines required for the execution of the plant life cycle

Subroutine 1	Cell Life
2	Embryonic development
3	How to be a seed
4	Bud development
5	Leaf development
6	Stem development
7	Root development
8	Reproductive development

So, we view the information needed for development as subdivided into categories, each category for a particular task. We next ask how might these subroutines be related to one another? Exactly how are they to be wired together to constitute a whole programme? Let us consider this problem by considering first a simpler example—one which confronts us however with the same difficult questions. The individual subroutines themselves provide us with the required example. Let us consider the subroutine for bud development. We cannot, of course, specify in detail all the items contained in that part of the genetic book which tells how to make buds. We can, however, at least guess some of the kinds of things which must necessarily, by the nature of bud development, be contained therein.

The bud arises from a single cell or a small group of cells, depending on the species. We choose, for simplicity, one in which a single cell produces by division, all the sub-apical cells. Thus, our apical cell must produce a bud and thence the bud-derived organs. It must have information about how to divide in appropriate planes. The different cell types of adult tissues differ in size, and our bud must therefore have information about when and where growth unattended by division is required. Thus one general category of information essential to the bud subroutine concerns cell division and growth. Clearly too, information must be provided about when and where to stop cell division. Such information might in some cases be merely a direction to divide so many times and no more. This cannot be true in the bud, however, since little pieces cut from the meristem of a bud continue cell division until a complete bud has been again produced. The same is true of the development of embryos from the isolated blastomeres of the early embryo. We need a new principle, and to this principle we assign the designation of test. In the bud, we shall imagine that the dividing and growing meristematic tissue continually tests itself for size or number of cells, each time comparing the value found with a value stated in the programme to be the desired one. When the correct value is attained, the cells of the bud can then proceed to the next developmental step. We shall have more to say later about the material nature of such tests. Let us now merely reflect on the concept itself for it carries us, I believe, to the central core of the logic of differentiation. The growing meristem tests itself against the desired ultimate size. The same bud tests for the presence or absence of flowering hormone so that it knows whether to turn into a flower bud or not. A cell tests its neighbour for strangeness or similarity. In these and a myriad of further ways must we imagine that each cell in the developing organism keeps itself informed of where it is and what, therefore, is the appropriate next step. The test as one unit, one component, of the logic of differentiation is, of course, already known as experimental fact. We know of some instances, such as those of the hormones and their target organs. All that we do here is to extend the concept to include other kinds of tests which, although not yet experimentally known, would seem to be essential to the machinery of development.

Finally, the subroutine for the development of the bud may, logically, include within itself further sub subroutines, pre-programmed sequences required, for example, to make the different kinds of specialized cells of the mature organ.

There are then three categories of instructions or commands, which are necessary components of a programme for bud development. These are, in summary, commands concerning cell division and growth, commands for the execution of various tests, and commands for the bringing into play, following the

TABLE 2

Programme for development of an organ from a single initial cell

Commands used

1	Divide tangentially with growth	
2	Divide transversely with growth	
3	Grow but do not divide	
4	Test for size or cell number which must achieve a specified value	
5	Test for apical or not apical	
6	Test for inside or not inside	
7	Test for outside or not outside	
8	Call Max	stored sub subroutines for differentiation into xylem and phloem respectively
9	Call Map	
10	Call Epidermis sub-subroutine	

proper outcome of the appropriate test, of an appropriate stored sub-subroutine. One set of commands appropriate for bud development which one might envisage are listed in Table 2.

It is interesting that the number of kinds of commands required for our purposes is so small—ten or eleven if we include the final command, 'Stop'. This number is small in part, of course, because we have masked many possible kinds of commands behind general headings as stored sub-subroutines. It is small too, because we have deliberately chosen to develop a bud which will not grow leaves. Even so, the insertion of two further commands, one a test and one a subroutine, would cause our bud to grow leaves too.

One way in which the commands of Table 2 might be sequenced to guide the apical cell of a growing bud along the pathway by which buds turn into stems is indicated in Fig. 1.

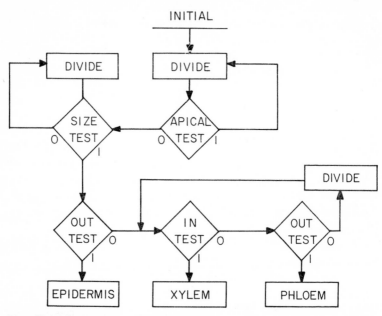

FIG. 12.1. An example of one way in which a set of genetic switching units (Table 2) might be interconnected to bring about the development of a single cell into an organ. In this example, a single apical cell generates a stem containing a variety of specialized cell types.

The logic of the arrangements in the programme is simple. The single apical cell divides transversely and each of the daughter cells then tests itself to find out whether it is still the apical one or not. For the apical cell, the result of the test is positive and the resulting command is to behave as an apical cell—that is divide again. The cell which has discovered that it is not apical receives a different command—namely to test for the size of the group of cells in which it is embedded and to compare this with the desired final size. The non-apical cell finds, as a result of the test, that it is smaller than the required final size. The resulting command is to divide some more—this time by, say, divisions in alternating planes, testing after each division, until the required size of the group of cells has been attained. When the required size has been reached, each cell is next required to test itself as to its position in the group. If a cell, as a result of this test, finds itself on the outside of the tissue mass, it is instructed to call upon the subroutine needed to generate epidermal cells. If, however, the outcome of the test is negative, the cell is required to test itself further to discover whether or not it is in the very centre of the tissue mass. If the outcome of this test is positive, the cell receives the command to call upon the subroutine required to generate the kinds of cells suitable for the inside of stems, namely xylem. A negative outcome requires a further test to determine whether or not the cell is on the outside of the undifferentiated tissue mass—namely next to the epidermis. This time a positive outcome requires calling upon the sub-subroutine for phloem generation. A negative outcome leads to a loop by which cells behave as cambium—divide and divide, each time testing to determine whether daughter cells shall turn into xylem or phloem.

The arrangements in Fig. 1 are all quite arbitrary. They do have the attractive feature that they accomplish the required task. To demonstrate that this is true, it is first necessary to write out the arrangements of Fig. 1 in more detail. This is done in Table 3. For the purposes of the programme of Table 3 we make two further simplifications. We will, firstly, ask our programme to generate a merely two-dimensional organ—a three-dimensional one would be quite possible but would complicate our task without revealing any new principles. We will, secondly, assume a special kind of time scale along which our bud progresses, a time scale subdivided into time units. In each time unit, each cell executes a single command—it divides, tests, or what-not. With these two provisos, let us see what structure the programme of Table 3 will generate. This is shown in Fig. 2. It generates first a growing bud and then, step by step, turns this into stem tissue surmounted, of course, by more growing bud. As long as the programme is followed, the apical cell will continue to divide and grow, leaving behind it an ever lengthening column of stem.

TABLE 3

Digital organ generator Model A (DOGMA)

Step	Command name	Command No.	Time unit
1	Divide transversely	2	1
2	Test for apical–not apical If 0, go to 3 If 1, go to 1	5	2
3	Divide tangentially	1	3
4	Size test, If 0, go to 5 If 1, go to 7	4	4
5	Divide transversely	2	5
6	Size test, If 0, go to 3 If 1, go to 7	4	6–12
7	Test for out–not out If 0, go to 10 If 1, go to 8	7	13
8	Grow	3	14
9	Make epidermis	10	15
10	Divide transversely	2	14
11	Test for in–not in If 0, go to 13 If 1, go to 12	6	15
12	Call Max	8	16
13	Test for out–not out If 0, go to 11 If 1, go to 14	7	17
14	Call Map	9	18

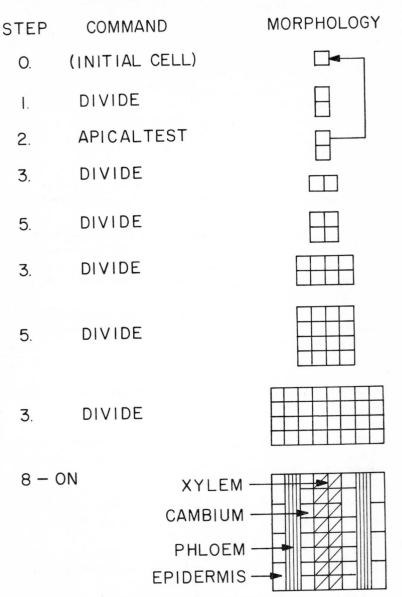

FIG. 12.2. The morphology generated by the execution of the programme of Table 3.

The exercise in which we have been indulging is not merely a game. It is an exercise which illuminates, to some degree, the nature of development. Development and differentiation is a complex task. We already know that nature, like man, accomplishes complex tasks by breaking them up into many simple sub-tasks. Our programme suggests the nature of some of the developmental sub-tasks. The progress of development requires that cells take appropriate, informed, action based upon the outcome of tests just as is done in carrying out any other complex task.

Although the scheme of Fig. 1 is utterly oversimplified, it nonetheless illustrates what appears to be inescapably the most basic of the principles involved in developmental matters, the concept of the test. The kinds of tests which we have envisaged and discussed are simple ones in which a single input determines one of only two alternative outputs. This is the kind of test to which our discussion of the hormones has already introduced us. Life may very well use more complicated tests, ones in which several inputs are weighed and processed to determine the output, the new posture of the genome. Let us, however, stick to our simple tests for the time being. What physical or chemical parameters might the genome utilize in the conduct of tests essential to development? The presence or absence in the nucleus of a particular hormone is, of course, one obvious example. Other kinds of tests are also known. Thus Moscona and Moscona (1963) have shown that certain cells of the chicken embryo continuously test adjacent cells for likeness or non-likeness. If neighbouring cells are alike, they stick together to form an organ. If neighbouring cells are different, they do not stick. The test is carried out by means of specific proteins of the cell surface. These sensor proteins are continuously degraded with a half-life of only a few minutes, but they are continuously replenished by ribosomal decoding of messenger RNA which, in turn, possesses a half-life of about 2 hours. Thus the recognition system of the embryo cells is under continuous genetic control.

One of the clearest examples of a developmental test at work is provided by the example of the plant embryo. The plant embryo develops, as do all embryos, from a single cell, normally the fertilized egg. This single cell is contained within an ovule, in which are the materials required for embryonic development. The cell from which the embryo develops need not be a fertilized egg. To cause a cell to develop into an embryo it is sufficient that two conditions be met. These are that the cell be single and that it be surrounded by the ovular (or endospermal) nutrients. Steward and his associates (1963) have shown that fully differentiated cells of many different types start life anew and develop into embryos normal for the species involved, provided only that these two conditions are fulfilled. We may imagine, in our present terms, that the plant cell is continuously testing itself for the presence of neighbours. If it finds that it has none, it then tests for the presence of the embryonic nutrients. If the outcome of this test is positive, the cell must say to itself, 'Well, it's strange, but I am all alone and here are the embryonic nutrients, so I must be a zygote and I will therefore develop into an embryo'. Incidentally, if the developing embryo is itself separated into single cells, each thinks this same thought; each starts anew as though it were a zygote.

We have used, in our earlier example, a test for apicalness. One physical basis upon which such a test might rest is cell contact. The cell could, for example, determine by its cellular recognition system, whether it is or is not in contact with other cells over say an angle of 180° or an angle of 270°. Similarly, we might envisage a size test as based upon recognition that some specified number of cells are now surrounded on all sides by like cells. The out–not out test which we have used could in principle be based upon the number of sides upon which a cell is surrounded by similar cells. In any case, there are vast numbers of physical and chemical parameters which are different for cells in different places in the developing cell mass. It is the task of the students of morphogenesis to determine which are significant to the tasks of development. We are here interested however in principle and the question of principle involved is: can tests of the kind which we envisage as required for development be conducted with the kind of hardware, the switching gear, which we now believe cells to have at their disposal? The answer is 'yes'. Our cell conducts a test in which it senses the concentration in the nucleus of a particular substance S_1. A positive outcome, that is a sufficiently high concentration of substance S_1, causes a deviation of course, the turning on of genes previously repressed. S_1 must then be an effector for those genes. The principle of cascade regulation makes possible further elaboration of the new course. What can the cell do after a negative outcome of the test, the sensing of a subthreshold concentration of substance S_1? It can leave matters as they are, continuing on its path until it meets another test. The basic switching element which we know the genetic material to possess is, in fact, exactly that required to conduct the tests which are the essential logic elements of the developmental programme.

We could, not readily to be sure, but in principle, rewrite the programme of Table 3 in a different language—that of effector, regulator, operator, and structural gene. We could similarly redraw the wiring diagram of Fig. 1 in terms of these same elements. To do so would, of course, make them more obscure to we users of human language, just as the machine language version of a computer programme is more obscure to us than is the human language version. We need have no doubt, however, that the effector, regulator, operator version of DOGMA (Digital Organ Generator Model A) would be more understandable to a cell than would be the human language version. It has been repeatedly discovered that analysis of a physical problem in terms specific and detailed enough to permit programming of that problem for a computer gives us deeper insight into the nature of the problem itself. The same may well be true of the problem of differentiation. It will be of interest to develop such programmes in much greater depth and detail than the example considered here. Such an effort could be based upon all relevant developmental anatomy and histology as well as upon any germane biochemical facts. The wealth of lore available from experimental embryology in particular should provide guide posts as to the proper format of the programme. And the programme, if it is to be useful, must suggest ways in which its validity can be tested; it must have predictive value. Since no detailed programme or model of a developmental process has as yet been made, we do not yet know with certainty if the approach will be a fruitful one. It does, however, appear to me that the approach to under-standing of the genetic-switching network by conceptual modelling holds promise; that by this approach we may hope in time to generate a sound general theory of the nature of development.

SELECTED REFERENCES

Flowering

ZEEVAART, J. A. D., *Science* **137**, 723 (1962).
ZEEVAART, J. A. D., in *Environmental Control of Plant Growth* (L. T. EVANS, Editor) p. 289. Academic Press (1963).
ZEEVAART, J. A. D., *Plant Physiol.* **37**, 296 (1962).

Developmental switching networks

MARUYAMA, M., *Am. Scient.* **50**, 164 (1962).

Tests

MOSCONA, M. H. and MOSCONA, A. A., *Science* **142**, 1070 (1963),
STEWARD, F. C., MAPES, M. O. and KENT, A. E., *Am. J. Bot.* **50**, 618 (1963).
STEWARD, F. C., *Sci. American.* **209**, 104 (1963).

SUMMARY AND PROSPECT

LET us now stand back and assess the view which we have developed, the view of differentiation as seen through the window of molecular biology. This view has been based wholly on the generally agreed-upon fact that development and differentiation consist in the orderly production by a single cell, the fertilized egg, of the several kinds of specialized cells which make up the adult creature. From this fact, we are inexorably led by logic to the conclusion that the observable cause of development and differentiation has its basis in a properly programmed expression and utilization of the genetic information. Thus, the different kinds of specialized cells differ from one another basically in that they contain different kinds of enzyme molecules. But enzyme molecules are specified by genes and all of the different kinds of specialized cells possess the same complement of genes. To study differentiation, therefore, we must and have considered what it is in the cell which determines that particular genes shall be active and elicit the enzyme for which they possess information and contrariwise, what it is that determines that particular genes shall be inactive. Genetic activity consists of transcription of the gene by RNA polymerase to yield the messenger RNA which carries the information of the gene. The final expression of genetic activity consists in the production by the ribosome of enzymes whose structure is determined by the information contained in the messenger RNA thus produced. The experimental study of differentiation as it is seen in this light requires, therefore, new tools and systems. We need to be able to isolate, in functional form, the genetic material of the cell. We require that the isolated genetic material possess, intact, the genetic controls characteristic of life. We require too, genetic material capable of the generation in the test tube of messenger RNA. We need ribosomal systems capable of decoding the messenger RNA thus produced so that we can determine what species of messenger RNA are produced by the isolated genetic material. These requirements have all been met. The methodology for the isolation of the genetic material is established. It has been shown, although only with respect to a small number of genes to be sure, that the genetic control characteristic of life is preserved in the isolated genome. We have discussed in detail the further methodology for the coupling of chromosomally-generated messenger RNA to messenger RNA-dependent ribosomal protein synthesizing systems.

We have then a system for the study, in the test tube, of the nature of the genetic control, which by repression or derepression of the individual elements of the genome, makes possible the course of development and differentiation. No doubt other and quite different kinds of systems for this type of study will be involved in future, but for the present, the kind of system which we are now considering offers a small infinity of opportunities for the gaining of further insight. We have discussed the control and programming of genetic activity on three levels, namely:

(1) the hardware of genetic control—the material nature of the repressor,

(2) the nature of the genetic switching unit—the nature of the act by which genetic activity is turned off or on, and

(3) the nature of the switching network by means of which the individual switching units are linked and integrated into a developmental system.

Concerning (1), the material nature of the repressor, something is now known. Some repressors are proteins both in higher creatures and in bacteria. In the chromosomes of higher creatures the repressor proteins include the histones although whether histones are the sole class of repressor proteins, we do not know. The logic by which repressor proteins discover the proper gene to repress remains to be uncovered, but these are matters now accessible to experimental study.

Concerning (2), the nature of the genetic switching unit, we also, as we have seen, begin to know something. Just as in bacteria so also in higher organisms particular kinds of small molecules possess the capability to turn off or on the activity of particular genes. In the bacteria, these small molecules and genes are concerned with everyday cellular metabolism, but the principle would appear to be the same in differentiation. Small molecules, the hormones for example, turn off or on individual or whole sets of genes in appropriate cells of higher organisms eliciting the production of characteristic enzyme molecules and, in appropriate instances, setting a cell or cells on a new pathway of development. This is most clearly and dramatically exemplified in the case of arousal from dormancy by ethylene chlorhydrin and induction of reproductive development by the flowering hormone. The same principle would, however, appear to be at work with a vast array of hormones, the estrogens and androgens, the cortico-steroids, gibberellic acid, the insect moulting hormone, ecdysone. What remains to be elucidated of course are the detailed molecular events of the derepression process—the way in which a hormone, by interaction with the genome, performs its derepressing function. But here again, the tools are at hand; the isolated genome in the repressed state, the hormone effector, the tools for analysis of whether or not derepression is effected. Analysis of the unit genetic switch will no doubt yield many surprises but we can at least rejoice that such analysis can commence.

This leads us to the final aspect of the biology of differentiation, the mapping of the developmental switching network. That the logic of development is based upon such a network, there can be no doubt. The fact, for example, that the several hormones, each itself acting upon a unit switch or switches, interact in their effects to elicit sequences of developmental processes, indicates at once that such interacting switching of genetic activity takes place in life. How can a genetic switching network be mapped, become known in molecular detail? The task will certainly be a vast one. One mode for such analysis is indicated by our considerations of the hormones. We could find out in detail what genes a particular hormone, say the flowering hormone, turns on or off. One effect of this initial switching act will be, we know, to make the cells of the ovary susceptible to the serial action of further hormones, kinetin, gibberellic acid, and auxin, as a result of which the ovary grows into a fruit. The initial switching act, performed by the flowering hormone, includes then sub-acts by which the posture of the genome of the ovarian cells is made receptive to the action of the hormones of fruit development and so, step by step, we may hope to outline individual portions of the switching network.

The points of control of genetic activity which we know or sense from the new understanding of the hormones are assuredly but the top of an iceberg. No doubt the genetic control points of which we have no inkling at all outnumber vastly those which we now begin to glimpse. Pontecorvo (1963) estimates that even *Aspergillus*, a relatively simple complex creature, may possess of the order of two regulator genes per structural gene and we may anticipate that in the still more complex creatures, the number will be far greater. It may well be then that the stochastic modelling of developmental switching networks along the lines proposed in the previous chapter can be of assistance in the solution of this complex task. Such modelling could, for example, assist us by suggesting the kinds of control points for which we might fruitfully look experimentally. It might indicate the logical necessity for the developmental process of control points, or control loops, which would otherwise long remain unsuspected and undetected.

It has been the object of this volume to assess the possibility and the potential fruitfulness of the study today of developmental processes on the level of molecular biology. The conclusion to which I have been lead—and hopefully the reader too, is that the time is ripe for such an attack. These are indeed stirring times. Never have the opportunities for the deep probing of the basic and innermost nature of developmental processes and systems been so great as they are today.

INDEX

ACTH, 122.
Actinomycin D, 59, 65, 88, 115, 122–6, 128, 129.
—, mode of action of, 120.
Alanine in histone, 82.
Alkaline phosphatase, 112.
Allosteric protein, 110.
Alpha-amylase, 128.
— gene, 115.
Alpha-helix, 98, 101.
Amino acid, 3.
— of histones, 80, 85.
Androgens, 123.
Annealing, DNA to RNA, 70.
—, DNA to histone, 92.
    See also Hybridization
Apical cell, division of, 135.
Arginine, 81, 97.
—, synthesis of, 107–8.
Aspartic acid, 101.
Aspergillus, 149.
ATP-ase, 111.
Autoradiography (*see* radioautography), 89.

Bacterial DNA, 42.
Bacteriophage, 37.
—, SP 8, 41.
—, ϕX-174, 38.
—, T4, 74, 105.
Barley plant, 115.
— seeds, 128.
Base complementarity, 37, 111.
Base composition of DNA and RNA, 37.
Base composition of RNA synthesised by polymerase, 37, 40, 126.
Base pairing, 44.
Base sequence, 71.
Benzpyrene hydroxylase, 125.
Bud development, 135.

Calf thymus, DNA-melting profile, 97.
Calf thymus histone, 79.
Canavanine, 112.
Carcinogens, 125.
Cell, 2.
— differentiation, 136.

Chicken, histones in organs of, 84.
Chicken embryo, organ development in, 142.
Chinese hamster, 89.
Chloroplasts, 2.
Chromatin, 113, 127, 129, 130.
— composition, 11, 26.
— detection, 34.
—, fractionation of, 12.
— function in nucleus, 65.
—, histone synthesis, 88.
—, isolation of, 7, 11, 35.
—, mechanical shearing of, 32.
—, melting profile of, 32.
— nucleohistone, 32.
— polymerase, 49.
—, properties of, 11, 33, 35.
—, proportion derepressed, 34, 101.
— in the ribosomal system, 49.
— in RNA synthesis, 8.
— sources, 33, 35.
— template, 21, 31, 33.
Chromatography of histones, 80.
Chromosome structure, 21, 127.
Chromosomes, DNA replication in, 43.
—, isolation and properties of, 7.
—, polytene, 43, 126.
—, ring structure in *E. coli*, 42.
Codon, 44.
Commands in differentiation, 137.
Cortisone, 117, 120.

Decoding of m-RNA, 46.
Dependency ratio, in ribosomal system, 48.
Deproteinisation of pea bud chromatin, 23.
Derepression (*see also* Repression), 24, 127, 129, 130, 133, 148.
— by ACTH, 122.
— by cortisone, 120.
— by hormones, 117.
Differentiation, 5, 136, 146.
DNA, aggregation, 92, 100.
— of chromatin, 41, 42.
— degradation of, 41.
— dependency ratio, 13, 30.